Applied Mathematics Research Cas

Kawsar Fatima

S.P Ahmad

Elias Munapo

Ahmed Buseri Ashine

978-1-62265-940-1 (online)

978-1-62265-941-8 (paper)

Published in United States

1-15 Bayesian Inference for Generalized Exponentiated Moment Exponential Distribution With Applications to Life Time Data by Kawsar Fatima and S.P Ahmad

16-31 POLYGON DEFORMATION APPROACH - A NEW FORMULATION FOR THE TRAVELING SALESMAN PROBLEM by Elias Munapo

32-36. Stability Analysis of Predatorprey model with ratio-dependent functional response by Ahmed Buseri Ashine

37-42. Predator-Prey Interactions With Desease In Predator Incorporating Harvesting Of Predator by Ahmed Buseri Ashine

978-1-62265-940-1 (online) 978-1-62265-941-8 (paper) - Applied Mathematics Research Cases

Bayesian Inference for Generalized Exponentiated Moment Exponential Distribution With Applications to Life Time Data

Kawsar Fatima and S.P Ahmad
Department of Statistics, University of Kashmir, Srinagar, India
Kawsar Fatima is Research Scholar in the Department of Statistics.
Email her at: kawsarfatima@gmail.com
S. P. Ahmad is an Assistant Professor in the Department of Statistics.
Email him at: sprvz@yahoo.com

Abstract: In this paper, we propose to obtain the Bayesian estimators of unknown shape parameter of a three parameter generalized exponentiated moment exponential (GEME) distribution, based on non-informative (Quasi and Extension of Jeffery's) priors using three different loss functions. Two real life data sets have been used to compare the performance of the estimates under different loss functions. The expression for survival function has also been established under Quasi prior and extension of Jeffrey's prior.

Keywords: GEME distribution, Baye's estimation, Priors, Survival function, Loss functions.

1 Introduction

The three-parameter generalized exponentiated moment exponential (GEME) distribution will be quite effectively used in analyzing several lifetime data, particularly in place of three-parameter gamma distribution, three parameter Weibull distribution or three-parameter exponentiated exponential distribution. Moment distributions have a vital role in mathematics and statistics, in particular in probability theory, in the perspective research related to ecology, reliability, biomedical field, econometrics, survey sampling and in life-testing. Hasnain (2013) developed an exponentiated moment exponential (EME) distribution and discussed some of its important properties. One of such distributions is the two-parameter weighted exponential distribution introduced by Gupta and Kundu (2009). Dara and Ahmad (2012) proposed a distribution function of moment exponential distribution and developed some basic properties like moments, skewness, kurtosis, moment generating function and hazard function. Bayes estimators for the weighted exponential distribution (WED) was considered by Farahani and Khorram (2014) while S.Dey et al. (2015) considered the estimation of the parameters of weighted exponential distribution. They obtained Bayes estimator for parameters of weighted exponential distribution by using weighted squared error loss function, modified squared error loss function, precautionary loss function, and K-loss function. They also compared the classical method with Bayesian method through simulation study. Recently, Devendera Kumar (2016) obtained the moments and estimation of the exponentiated moment exponential distribution. They obtained Bayes estimation under symmetric and asymmetric loss functions using gamma prior for both shape and scale parameters. They also compared the classical method with Bayesian method through Monte Carlo simulation.

As given in Zafar Iqbal et *al* (2014), the cumulative distribution function (cdf) of three parameter generalized exponentiated moment exponential (GEME) distribution is given by

$$F(x) = \left[1 - \frac{x^\gamma + \beta}{\beta} e^{\frac{-x^\gamma}{\beta}} \right]^\alpha, \quad x > 0 \tag{1.1}$$

Where α, β and γ are positive real parameters. The probability density function (pdf) of GEME distribution is defined as

$$f(x) = \frac{\alpha\gamma}{\beta^2} \left[1 - \left(1 + \frac{x^\gamma}{\beta} \right) e^{\frac{-x^\gamma}{\beta}} \right]^{\alpha-1} x^{2\gamma-1} e^{\frac{-x^\gamma}{\beta}}, \quad x > 0, \alpha, \beta, \gamma > 0 \tag{1.2}$$

Here α and γ are the shape parameters and β is the scale parameter. For $\gamma = 1$, it represents the EME distribution, for $\alpha = \gamma = 1$, it represents the size biased moment exponential distribution and for $\beta = \gamma = 1$, it represents the one parameter exponentiated exponential distribution.

2 Survival Function

The branch of statistics that deals with the failure in mechanical systems is called survival analysis. In engineering, it is called reliability analysis or reliability theory. In fact the survival function is the probability of failure by time y, where y represents survival time. We use survival function to predict quantiles of the survival time. Survival function, by definition, is

$$S(x) = 1 - F(x) = 1 - \left[1 - \left(1 + \frac{x^\gamma}{\beta} \right) e^{\frac{-x^\gamma}{\beta}} \right]^\alpha, \quad x > 0, \alpha, \beta, \gamma > 0 \tag{2.1}$$

and the hazard function is

$$h(x) = \frac{\dfrac{\alpha\gamma}{\beta^2} \left[1 - \left(1 + \dfrac{x^\gamma}{\beta} \right) e^{\frac{-x^\gamma}{\beta}} \right]^{\alpha-1} x^{2\gamma-1} e^{\frac{-x^\gamma}{\beta}}}{1 - \left[1 - \left(1 + \dfrac{x^\gamma}{\beta} \right) e^{\frac{-x^\gamma}{\beta}} \right]^\alpha}; \quad 0 < x < \infty, \alpha, \beta, \gamma > 0 \tag{2.2}$$

3. Maximum likelihood Estimation of Generalized Exponentiated Moment Exponential (GEME) distribution the shape parameter α is unknown:

Theorem 3.1: - Let $\underline{x} = (x_1, x_2, ..., x_n)$ be a random sample of size n having pdf (1.2); then the maximum likelihood estimator of shape parameter α, when the parameters γ and β are known, is given by

$$\hat{\alpha}_{MLE} = \frac{n}{\sum_{i=1}^{n} \ln\left[1-\left(1+\frac{x_i^{\gamma}}{\beta}\right)e^{\frac{-x_i^{\gamma}}{\beta}}\right]^{-1}} \qquad (3.1)$$

Proof: - The likelihood function of the pdf (1.2) is given by

$$L(\underline{x}/\alpha) = \frac{\alpha^n \gamma^n}{\beta^{2n}} \prod_{i=1}^{n}\left[1-\left(1+\frac{x_i^{\gamma}}{\beta}\right)e^{\frac{-x_i^{\gamma}}{\beta}}\right]^{\alpha-1} x_i^{2\gamma-1} e^{\frac{-x_i^{\gamma}}{\beta}} \qquad (3.2)$$

The log likelihood function is given by

$$\ln L(\underline{x}/\alpha) = n\ln\alpha + n\ln\gamma - 2n\ln\beta + (2\gamma-1)\sum_{i=1}^{n}\ln x_i - \frac{\sum_{i=1}^{n} x_i^{\gamma}}{\beta} + (\alpha-1)\sum_{i=1}^{n}\ln\left[1-\left(1+\frac{x_i^{\gamma}}{\beta}\right)e^{\frac{-x_i^{\gamma}}{\beta}}\right] \qquad (3.3)$$

Differentiating (3.3) with respect to α and equating to zero, we get

$$\frac{\partial \ln L(\underline{x}/\alpha)}{\partial \alpha} = \frac{n}{\alpha} + \sum_{i=1}^{n}\ln\left[1-\left(1+\frac{x_i^{\gamma}}{\beta}\right)e^{\frac{-x_i^{\gamma}}{\beta}}\right] = 0$$

$$\hat{\alpha}_{MLE} = \frac{n}{\sum_{i=1}^{n} \ln\left[1-\left(1+\frac{x_i^{\gamma}}{\beta}\right)e^{\frac{-x_i^{\gamma}}{\beta}}\right]^{-1}} \qquad (3.4)$$

4 Bayes Estimator: we now derive the Bayes estimator of the shape parameter α in GEMED when the parameters β and γ are assumed to be known. We consider two different priors and three different loss functions.

a) Quasi Prior: when there is no information about the parameter α, one may use the quasi density as given by:

$$g_1(\alpha) \propto \frac{1}{\alpha^d}, \qquad \alpha > 0, \ d > 0$$

The quasi prior leads to diffuse prior when d=0 and to a non informative prior for a case when d=1.

b) Extension of Jeffery's Prior: The extended Jeffrey's prior proposed by Al-Kutubi (2005), is given as:

$$g_2(\alpha) \propto [I(\alpha)]^{c_1}, c_1 \in R^+$$

$$[I(\alpha)] = -nE\left[\frac{\partial^2}{\partial \alpha^2} \log f(x)\right]$$ is the Fisher's information matrix. For the model (1.2),

$$g_2(\alpha) = k\left[\frac{n}{\alpha^2}\right]^{c_1} \Rightarrow g_2(\alpha) \propto \frac{1}{\alpha^{2c_1}}$$

i) Squared Error loss function

A commonly used loss function is the square error loss function (SELF)

$$l(\hat{\alpha}, \alpha) = c(\hat{\alpha} - \alpha)^2$$

which is symmetric loss function that assigns equal losses to over estimation and under estimation. The SELF is often used because it does not need extensive numerical computation.

a) Entropy Loss Function

In many practical situations, it appears to be more realistic to express the loss in terms of the ratio $\frac{\hat{\alpha}}{\alpha}$. In this case, Calabria and Pulcini (1994) point out that a useful asymmetric loss function is the entropy loss function $L(\delta^p) \propto [\delta^p - p\log(\delta) - 1]$ where $\delta = \frac{\hat{\alpha}}{\alpha}$ and $p>0$, whose minimum occur at $\hat{\alpha} = \alpha$. Also, the loss function $L(\delta)$ has been used in Dey et al (1987) and Dey and Liu (1992), in the original form having $p = 1$. Thus, $L(\delta)$ can be written as

$$L(\delta) = b_1[\delta - \log(\delta) - 1] \; ; b_1 > 0.$$

iii) Al-Bayyati's loss function

Al-Bayyati's (2002) introduced a new loss function of the form $l(\hat{\alpha}, \alpha) = \alpha^{c_2}(\hat{\alpha} - \alpha)^2 \; ; c_2 \in R^+$. Which is symmetric α and $\hat{\alpha}$ represent the true and estimated values of the parameter. This loss function is frequently used because of its analytical tractability in Bayesian analysis.

4.1 Posterior density under Quasi Prior:

Under quasi prior, using (3.1), the posterior distribution of parameter α is given by

$$p(\alpha/\underline{x}) \propto \frac{\alpha^n \gamma^n}{\beta^{2n}} \prod_{i=1}^{n} \left[1 - \left(1 + \frac{x_i^\gamma}{\beta}\right) e^{\frac{-x_i^\gamma}{\beta}}\right]^{\alpha-1} x_i^{2\gamma-1} e^{\frac{-x_i^\gamma}{\beta}} \frac{1}{\alpha^d}$$

$$p(\alpha / \underline{x}) = k\alpha^{n-d} e^{-\alpha \sum_{i=1}^{n} \ln\left[1-\left(1+\frac{x_i^{\gamma}}{\beta}\right)e^{\frac{-x_i^{\gamma}}{\beta}}\right]^{-1}}$$

$$p(\alpha / \underline{x}) = k\alpha^{n-d} e^{-\alpha\beta_1} \tag{4.1}$$

where k is independent of α, $\beta_1 = \sum_{i=1}^{n} \ln\left[1-\left(1+\frac{x_i^{\gamma}}{\beta}\right)e^{\frac{-x_i^{\gamma}}{\beta}}\right]^{-1}$ and $k^{-1} = \int_{0}^{\infty}\alpha^{n-d}e^{-\alpha\beta_1}\,d\alpha$

$$\Rightarrow k^{-1} = \frac{\Gamma(n-d+1)}{\beta_1^{n-d+1}}$$

Therefore from (4.1) we have

$$p_{1Q}(\alpha / \underline{x}) = \frac{\beta_1^{n-d+1}}{\Gamma(n-d+1)}\alpha^{n-d}e^{-\alpha\beta_1}\quad,\quad \alpha > 0 \tag{4.2}$$

which is the density kernel of gamma distribution having parameters $\alpha_1 = (n-d+1) and$

$\beta_1 = \sum_{i=1}^{n} \ln\left[1-\left(1+\frac{x_i^{\gamma}}{\beta}\right)e^{\frac{-x_i^{\gamma}}{\beta}}\right]^{-1}$. So the posterior distribution of $(\alpha / \underline{x}) \sim G(\alpha_1,\beta_1)$

4.2 Bayesian estimation by using Quasi prior under different Loss Functions:

Theorem 4.2.1:- Assuming the loss function $l_{1QS}(\hat{\alpha},\alpha)$, the Bayesian estimator of the shape parameter α, when the parameters β and γ are assumed to be known, is of the form

$$\hat{\alpha}_{1QS} = \frac{(n-d+1)}{\beta_1}\quad ; \beta_1 = \sum_{i=1}^{n} \ln\left[1-\left(1+\frac{x_i^{\gamma}}{\beta}\right)e^{\frac{-x_i^{\gamma}}{\beta}}\right]^{-1}$$

Proof: - The risk function of the estimator α under the squared error loss function $L_{1QS}(\hat{\alpha},\alpha)$ is given by the formula

$$R(\hat{\alpha},\alpha) = \int_{0}^{\infty} c(\hat{\alpha}-\alpha)^2 p_{1Q}(\alpha / \underline{x})\,d\alpha \tag{4.3}$$

On substituting (4.2) in (4.3), we have

$$R(\hat{\alpha}, \alpha) = \int_0^\infty c(\hat{\alpha} - \alpha)^2 \frac{\beta_1^{n-d+1}}{\Gamma(n-d+1)} \alpha^{n-d} e^{-\alpha\beta_1} \, d\alpha$$

$$R(\hat{\alpha}, \alpha) = c \frac{\beta_1^{n-d+1}}{\Gamma(n-d+1)} \left[\hat{\alpha}^2 \int_0^\infty \alpha^{n-d+1-1} e^{-\alpha\beta_1} d\alpha + \int_0^\infty \alpha^{n-d+3-1} e^{-\alpha\beta_1} d\alpha - 2\hat{\alpha} \int_0^\infty \alpha^{n-d+2-1} e^{-\alpha\beta_1} d\alpha \right]$$

$$(4.4)$$

On solving (4.4), we get

$$R(\hat{\alpha}, \alpha) = c \left[\hat{\alpha}^2 + \frac{(n-d+2)(n-d+1)}{\beta_1^2} - 2\hat{\alpha} \frac{(n-d+1)}{\beta_1} \right]$$

Minimization of the risk with respect to $\hat{\alpha}$ gives us the optimal estimator

$$\hat{\alpha}_{1QS} = \frac{(n-d+1)}{\beta_1} \quad ; \beta_1 = \sum_{i=1}^n \ln \left[1 - \left(1 + \frac{x_i^\gamma}{\beta} \right) e^{\frac{-x_i^\gamma}{\beta}} \right]^{-1}$$

$$(4.5)$$

Theorem 4.2.2:- Assuming the loss function $l_{1QE}(\hat{\alpha}, \alpha)$, the Bayesian estimator of the shape parameter α, when the parameters β and γ are assumed to be known, is of the form

$$\hat{\alpha}_{1QE} = \frac{(n-d)}{\beta_1} \quad ; \beta_1 = \sum_{i=1}^n \ln \left[1 - \left(1 + \frac{x_i^\gamma}{\beta} \right) e^{\frac{-x_i^\gamma}{\beta}} \right]^{-1}$$

Proof: - The risk function of the estimator α under the entropy loss function $l_{1QE}(\hat{\alpha}, \alpha)$ is given by the formula

$$R(\hat{\alpha}, \alpha) = \int_0^\infty b_1 \left(\frac{\hat{\alpha}}{\alpha} - \log\left(\frac{\hat{\alpha}}{\alpha} \right) - 1 \right) p_{1Q}(\alpha / \underline{x}) \, d\alpha$$

$$(4.6)$$

On substituting (4.2) in (4.6), we have

$$R(\hat{\alpha}, \alpha) = \int_0^\infty b_1 \left(\frac{\hat{\alpha}}{\alpha} - \log\left(\frac{\hat{\alpha}}{\alpha} \right) - 1 \right) \frac{\beta_1^{n-d+1}}{\Gamma(n-d+1)} \alpha^{n-d} e^{-\alpha\beta_1} \, d\alpha$$

$$(4.7)$$

On solving (4.7), we get

$$R(\hat{\alpha}, \alpha) = b_1 \left[\frac{\hat{\alpha}\beta_1}{(n-d)} - \log(\hat{\alpha}) + \frac{\Gamma'(n-d+1)}{\Gamma(n-d+1)} - 1 \right]$$

Minimization of the risk with respect to $\hat{\alpha}$ gives us the optimal estimator

$$\hat{\alpha}_{1QE} = \frac{(n-d)}{\beta_1} \quad \hat{\alpha}_{1QE} = \frac{(n-d)}{\beta_1} \quad ;\beta_1 = \sum_{i=1}^{n} \ln\left[1-\left(1+\frac{x_i^\gamma}{\beta}\right)e^{\frac{-x_i^\gamma}{\beta}}\right]^{-1} \quad (4.8)$$

Theorem 4.2.3:- Assuming the loss function $l_{1QA}(\hat{\alpha},\alpha)$, the Bayesian estimator of the shape parameter α, when the parameters β and γ are assumed to be known, is of the form

$$\hat{\alpha}_{1QA} = \frac{(c_2+n-d+1)}{\beta_1} \quad ;\beta_1 = \sum_{i=1}^{n} \ln\left[1-\left(1+\frac{x_i^\gamma}{\beta}\right)e^{\frac{-x_i^\gamma}{\beta}}\right]^{-1}$$

Proof: - The risk function of the estimator α under the Al-Bayyati loss function $l_{1QA}(\hat{\alpha},\alpha)$ is given by the formula

$$R(\hat{\alpha},\alpha) = \int_0^\infty \alpha^{c2}(\hat{\alpha}-\alpha)^2 p_{1Q}(\alpha/\underline{x})\ d\alpha \quad (4.9)$$

On substituting (4.2) in (4.9), we have

$$R(\hat{\alpha},\alpha) = \int_0^\infty \alpha^{c2}(\hat{\alpha}-\alpha)^2 \frac{\beta_1^{n-d+1}}{\Gamma(n-d+1)} \alpha^{n-d} e^{-\alpha\beta_1}\ d\alpha \quad (4.10)$$

On solving (4.10), we get

$$R(\hat{\alpha},\alpha) = \frac{1}{\Gamma(n-d+1)}\left[\hat{\alpha}^2 \frac{\Gamma(c_2+n-d+1)}{\beta_1^{c2}} + \frac{\Gamma(c_2+n-d+3)}{\beta_1^{c2+2}} - 2\hat{\alpha}\frac{\Gamma(c_2+n-d+2)}{\beta_1^{c2+1}}\right]$$

Minimization of the risk with respect to $\hat{\alpha}$ gives us the optimal estimator

$$\hat{\alpha}_{1QA} = \frac{(c_2+n-d+1)}{\beta_1} \quad ;\beta_1 = \sum_{i=1}^{n} \ln\left[1-\left(1+\frac{x_i^\gamma}{\beta}\right)e^{\frac{-x_i^\gamma}{\beta}}\right]^{-1} \quad (4.11)$$

Remark 1.1 Replacing $c_2=0$, in (4.11), we get the Baye's estimator under square error loss function with quasi prior which is same as (4.5)

4.3 Posterior density under Extension of Jeffery's prior
Under extension of Jeffery's prior, using (3.1), the posterior distribution of the unknown parameter α is given by

$$p_{2E}(\alpha/\underline{x}) \propto \frac{\alpha^n \gamma^n}{\beta^{2n}} \prod_{i=1}^{n}\left[1-\left(1+\frac{x_i^\gamma}{\beta}\right)e^{\frac{-x_i^\gamma}{\beta}}\right]^{\alpha-1} x_i^{2\gamma-1} e^{\frac{-x_i^\gamma}{\beta}} \frac{1}{\alpha^{2c_1}}$$

$$p_{2E}(\alpha/\underline{x})= k\alpha^{n-2c_1} e^{-\alpha \sum_{i=1}^{n} \ln\left[1-\left(1+\frac{x_i^\gamma}{\beta}\right)e^{\frac{-x_i^\gamma}{\beta}}\right]^{-1}}$$

$$p_{2E}(\alpha/\underline{x})= k\alpha^{n-2c_1} e^{-\alpha\beta_1} \tag{4.12}$$

where k is independent of α, $\beta_1 = \sum_{i=1}^{n} \ln\left[1-\left(1+\frac{x_i^\gamma}{\beta}\right)e^{\frac{-x_i^\gamma}{\beta}}\right]^{-1}$ and $k^{-1} = \int_0^\infty \alpha^{n-2c_1} e^{-\alpha\beta_1} d\alpha$

$$\Rightarrow k^{-1} = \frac{\Gamma(n-2c_1+1)}{\beta_1^{n-2c_1+1}}$$

Therefore from (4.12) we have

$$p_{2E}(\alpha/\underline{x})= \frac{\beta_1^{n-2c_1+1}}{\Gamma(n-2c_1+1)} \alpha^{n-2c_1} e^{-\alpha\beta_1} \quad , \alpha > 0 \tag{4.13}$$

which is the density kernel of gamma distribution having parameters $\alpha_1 = (n-2c_1+1)$ and

$\beta_1 = \sum_{i=1}^{n} \ln\left[1-\left(1+\frac{x_i^\gamma}{\beta}\right)e^{\frac{-x_i^\gamma}{\beta}}\right]^{-1}$. So the posterior distribution of $(\alpha/\underline{x}) \sim G(\alpha_2, \beta_1)$

4.4 Bayesian estimation by using Extension of prior under different Loss Functions:

Theorem 4.4.1:- Assuming the loss function $l_{2EJS}(\hat{\alpha}, \alpha)$, the Bayesian estimator of the shape parameter α, when the parameters β and γ are assumed to be known, is of the form

$$\hat{\alpha}_{2EJS} = \frac{(n-2c_1+1)}{\beta_1} \quad ; \beta_1 = \sum_{i=1}^{n} \ln\left[1-\left(1+\frac{x_i^\gamma}{\beta}\right)e^{\frac{-x_i^\gamma}{\beta}}\right]^{-1}$$

Proof: - The risk function of the estimator α under the squared error loss function $l_{2EJS}(\hat{\alpha}, \alpha)$ is given by the formula

$$R(\hat{\alpha}, \alpha) = \int_0^\infty c(\hat{\alpha}-\alpha)^2 p_{2E}(\alpha/\underline{x}) \, d\alpha \tag{4.14}$$

On substituting (4.13) in (4.14), we have

$$R(\hat{\alpha},\alpha) = \int_0^\infty c(\hat{\alpha}-\alpha)^2 \frac{\beta_1^{n-2c_1+1}}{\Gamma(n-2c_1+1)} \alpha^{n-2c_1} e^{-\alpha\beta_1} \, d\alpha \qquad (4.15)$$

On solving (4.15), we get

$$R(\hat{\alpha},\alpha) = c\left[\hat{\alpha}^2 + \frac{(n-2c_1+2)(n-2c_1+1)}{\beta_1^2} - 2\hat{\alpha}\frac{(n-2c_1+1)}{\beta_1} \right]$$

Minimization of the risk with respect to $\hat{\alpha}$ gives us the optimal estimator

$$\hat{\alpha}_{2EJS} = \frac{(n-2c_1+1)}{\beta_1} \quad ; \beta_1 = \sum_{i=1}^n \ln\left[1 - \left(1+\frac{x_i^{\gamma}}{\beta}\right) e^{\frac{-x_i^{\gamma}}{\beta}} \right]^{-1} \qquad (4.16)$$

Theorem 4.4.2:- Assuming the loss function $l_{2EJE}(\hat{\alpha},\alpha)$, the Bayesian estimator of the shape parameter α, when the parameters β and γ are assumed to be known, is of the form

$$\hat{\alpha}_{2EJE} = \frac{(n-2c_1)}{\beta_1} \quad ; \beta_1 = \sum_{i=1}^n \ln\left[1 - \left(1+\frac{x_i^{\gamma}}{\beta}\right) e^{\frac{-x_i^{\gamma}}{\beta}} \right]^{-1}$$

Proof: - The risk function of the estimator α under the entropy loss function $l_{2EJE}(\hat{\alpha},\alpha)$ is given by the formula

$$R(\hat{\alpha},\alpha) = \int_0^\infty b_1\left(\frac{\hat{\alpha}}{\alpha} - \log\left(\frac{\hat{\alpha}}{\alpha}\right) - 1 \right) p_{2E}(\alpha/\underline{x}) \, d\alpha \qquad (4.17)$$

On substituting (4.13) in (4.17), we have

$$R(\hat{\alpha},\alpha) = \int_0^\infty b_1\left(\frac{\hat{\alpha}}{\alpha} - \log\left(\frac{\hat{\alpha}}{\alpha}\right) - 1 \right) \frac{\beta_1^{n-2c_1+1}}{\Gamma(n-2c_1+1)} \alpha^{n-2c_1} e^{-\alpha\beta_1} \, d\alpha \qquad (4.18)$$

On solving (4.18), we get

$$R(\hat{\alpha},\alpha) = b_1\left[\frac{\hat{\alpha}\beta_1}{(n-2c_1)} - \log(\hat{\alpha}) + \frac{\Gamma'(n-2c_1+1)}{\Gamma(n-2c_1+1)} - 1 \right]$$

Minimization of the risk with respect to $\hat{\alpha}$ gives us the optimal estimator

$$\hat{\alpha}_{2EJE} = \frac{(n-2c_1)}{\beta_1} \quad ; \beta_1 = \sum_{i=1}^{n} \ln\left[1-\left(1+\frac{x_i^{\gamma}}{\beta}\right)e^{\frac{-x_i^{\gamma}}{\beta}}\right]^{-1} \tag{4.19}$$

Theorem 4.4.3:- Assuming the loss function $l_{2EJA}(\hat{\alpha}, \alpha)$, the Bayesian estimator of the shape parameter α, when the parameters β and γ are assumed to be known, is of the form

$$\hat{\alpha}_{2EJA} = \frac{(c_2+n-2c_1+1)}{\beta_1} \quad ; \beta_1 = \sum_{i=1}^{n} \ln\left[1-\left(1+\frac{x_i^{\gamma}}{\beta}\right)e^{\frac{-x_i^{\gamma}}{\beta}}\right]^{-1}$$

Proof: - The risk function of the estimator α under the entropy loss function $l_{2EJA}(\hat{\alpha}, \alpha)$ is given by the formula

$$R(\hat{\alpha}, \alpha) = \int_{0}^{\infty} \alpha^{c2}(\hat{\alpha}-\alpha)^2 p_{2E}(\alpha/\underline{x}) \; d\alpha \tag{4.20}$$

On substituting (4.13) in (4.20), we have

$$R(\hat{\alpha}, \alpha) = \int_{0}^{\infty} \alpha^{c2}(\hat{\alpha}-\alpha)^2 \frac{\beta_1^{n-2c_1+1}}{\Gamma(n-2c_1+1)} \alpha^{n-2c_1} e^{-\alpha\beta_1} \; d\alpha \tag{4.21}$$

On solving (4.21), we get

$$R(\hat{\alpha}, \alpha) = \frac{1}{\Gamma(n-2c_1+1)}\left[\hat{\alpha}^2 \frac{\Gamma(c_2+n-2c_1+1)}{\beta_1^{c2}} + \frac{\Gamma(c_2+n-2c_1+3)}{\beta_1^{c2+2}} - 2\hat{\alpha}\frac{\Gamma(c_2+n-2c_1+2)}{\beta_1^{c2+1}}\right]$$

Minimization of the risk with respect to $\hat{\alpha}$ gives us the optimal estimator

$$\hat{\alpha}_{2EJA} = \frac{(c_2+n-2c_1+1)}{\beta_1} \quad ; \beta_1 = \sum_{i=1}^{n} \ln\left[1-\left(1+\frac{x_i^{\gamma}}{\beta}\right)e^{\frac{-x_i^{\gamma}}{\beta}}\right]^{-1} \tag{4.22}$$

5. Estimator of survival Function

5.1 Estimator under Quasi prior of survival Function

By using posterior distribution function, we can found the survival function such that

$$\hat{S}_{1Q}(x) = \int_{0}^{\infty}\left\{1-\left(1-\left(1+\frac{x^{\gamma}}{\beta}\right)e^{\frac{-x^{\gamma}}{\beta}}\right)^{\alpha}\right\}p_{1Q}(\alpha/\underline{x})d\alpha$$

$$\hat{S}_{1Q}(x) = \int_0^\infty \left\{ 1 - \left(1 - \left(1 + \frac{x^\gamma}{\beta} \right) e^{\frac{-x^\gamma}{\beta}} \right)^\alpha \right\} \frac{\beta_1^{n-d+1}}{\Gamma(n-d+1)} \alpha^{n-d} e^{-\alpha\beta_1} d\alpha$$

$$\hat{S}_{1Q}(x) = \int_0^\infty \frac{\beta_1^{n-d+1}}{\Gamma(n-d+1)} \alpha^{n-d} e^{-\alpha\beta_1} d\alpha - \int_0^\infty \left(1 - \left(1 + \frac{x^\gamma}{\beta} \right) e^{\frac{-x^\gamma}{\beta}} \right)^\alpha \frac{\beta_1^{n-d+1}}{\Gamma(n-d+1)} \alpha^{n-d} e^{-\alpha\beta_1} d\alpha$$

$$\hat{S}_{1Q}(x) = 1 - \left(\frac{\beta_1}{\beta_1 - \beta_2} \right)^{n-d+1} \tag{5.1}$$

where $\beta_1 = \sum_{i=1}^n \ln\left[1 - \left(1 + \frac{x_i^\gamma}{\beta} \right) e^{\frac{-x_i^\gamma}{\beta}} \right]^{-1}$ and $\beta_2 = \ln\left[1 - \left(1 + \frac{x^\gamma}{\beta} \right) e^{\frac{-x^\gamma}{\beta}} \right]$

5.2 Estimator under Extension prior of survival Function

By using posterior distribution function, we can found the survival function such that

$$\hat{S}_{2E}(x) = \int_0^\infty \left\{ 1 - \left(1 - \left(1 + \frac{x^\gamma}{\beta} \right) e^{\frac{-x^\gamma}{\beta}} \right)^\alpha \right\} p_{2E}(\alpha / \underline{x}) d\alpha$$

$$\hat{S}_{2E}(x) = \int_0^\infty \left\{ 1 - \left(1 - \left(1 + \frac{x^\gamma}{\beta} \right) e^{\frac{-x^\gamma}{\beta}} \right)^\alpha \right\} \frac{\beta_1^{n-2c_1+1}}{\Gamma(n-2c_1+1)} \alpha^{n-2c_1} e^{-\alpha\beta_1} d\alpha$$

$$\hat{S}_{2E}(x) = \int_0^\infty \frac{\beta_1^{n-2c_1+1}}{\Gamma(n-2c_1+1)} \alpha^{n-2c_1} e^{-\alpha\beta_1} d\alpha - \int_0^\infty \left(1 - \left(1 + \frac{x^\gamma}{\beta} \right) e^{\frac{-x^\gamma}{\beta}} \right)^\alpha \frac{\beta_1^{n-2c_1+1}}{\Gamma(n-2c_1+1)} \alpha^{n-2c_1} e^{-\alpha\beta_1} d\alpha$$

$$\hat{S}_{2E}(x) = 1 - \left(\frac{\beta_1}{\beta_1 - \beta_2} \right)^{n-2c_1+1} \tag{5.2}$$

where $\beta_1 = \sum_{i=1}^n \ln\left[1 - \left(1 + \frac{x_i^\gamma}{\beta} \right) e^{\frac{-x_i^\gamma}{\beta}} \right]^{-1}$ and $\beta_2 = \ln\left[1 - \left(1 + \frac{x^\gamma}{\beta} \right) e^{\frac{-x^\gamma}{\beta}} \right]$

6. Applications

To compare the performance of the estimates under different loss functions for the generalized exponentiated moment exponential distribution, two real data sets are used and analysis performed with the help of R software.

Data set I: The first data set is given by Gross and Clark (1975, P. 105) which represents the lifetime's data relating to relief times (in minutes) of 20 patients receiving an analgesic. The data are as follows: 1.1, 1.4, 1.3, 1.7, 1.9, 1.8, 1.6, 2.2, 1.7, 2.7, 4.1, 1.8, 1.5, 1.2, 1.4, 3.0, 1.7, 2.3, 1.6, and 2.0

Table 1: Bayes Risk of α under Quasi prior for Data set I

β	γ	d	SELF		ELF		ABLF	
			$c = 0.5$	$c = 1.0$	$b_1 = 0.2$	$b_1 = 0.4$	$c_2 = 0.5$	$c_2 = -0.5$
1.0	1.5	0.3	0.12806	0.25611	0.44425	0.88851	0.39562	0.16777
		1.3	0.12187	0.24374	0.44452	0.88904	0.36763	0.16361
1.5	2.0	0.3	0.07248	0.14496	0.50117	1.00234	0.19423	0.10948
		1.3	0.06898	0.13796	0.50143	1.00287	0.18049	0.10677
2.0	2.5	0.3	0.05948	0.11897	0.52092	1.04185	0.15172	0.09440
		1.3	0.05661	0.11323	0.52119	1.04239	0.14099	0.09206
2.5	3.0	0.3	0.05711	0.11422	0.52501	1.05001	0.14418	0.09155
		1.3	0.05435	0.10871	0.52527	1.05054	0.13398	0.08929

SELF: squared error loss function, ELF: entropy loss function, and ABLF: Al-Bayyati's loss function.

Table 2: Bayes Risk of α under Extension of Jeffery's prior for Data set I

β	γ	c_1	SELF		ELF		ABLF	
			$c = 0.5$	$c = 1.0$	$b_1 = 0.2$	$b_1 = 0.4$	$c_2 = 0.5$	$c_2 = -0.5$
1.0	1.5	0.4	0.12496	0.24993	0.44438	0.88877	0.38154	0.16571
		1.4	0.11259	0.22518	0.44498	0.88996	0.32695	0.15718
1.5	2.0	0.4	0.07073	0.14146	0.50130	1.00260	0.18732	0.10813
		1.4	0.06373	0.12745	0.50189	1.00379	0.16051	0.10257
2.0	2.5	0.4	0.05805	0.11610	0.52105	1.04211	0.146326	0.09324
		1.4	0.05230	0.10461	0.52165	1.04331	0.12538	0.08844
2.5	3.0	0.4	0.055732	0.11146	0.52513	1.05027	0.13905	0.09043
		1.4	0.05021	0.10043	0.52573	1.05146	0.11916	0.08578

SELF: squared error loss function, ELF: Entropy loss function and ABLF: Al-Bayyati's loss function.

From Table 1 and 2 shows that squared error loss function provides the minimum posterior risk as compared to the other loss functions particularly as C is (0.5) and the prior Extension of Jeffery's prior provides the less posterior risk than Quasi prior. Moreover, when we increase the true value of parameters $(\beta, \gamma) = c((1.0, 1.5), (1.5, 2.0)$ and $(2.5, 3.0))$ and increase the value of $d = c(0.3, 1.3)$ and $c_1 = c(0.4, 1.4)$, the Bayes risk of $\hat{\alpha}$ decreases quite significantly.

Data set II: The second data set studied by Meeker and Escobar (1998), which gives the times of failure and running times for a sample of devices from a eld-tracking study of a larger system. At a certain point in time, 30 units were installed in normal service conditions. Two causes of failure were observed for each unit that failed: the failure caused by an accumulation of randomly occurring damage from power-line voltage spikes during electric storms and failure caused by normal product wear. The times are:

2.75, 0.13, 1.47, 0.23, 1.81, 0.30, 0.65, 0.10, 3.00, 1.73, 1.06, 3.00, 3.00, 2.12, 3.00, 3.00, 3.00, 0.02, 2.61, 2.93, 0.88, 2.47, 0.28, 1.43, 3.00, 0.23, 3.00, 0.80, 2.45, 2.66.

Table 3: Bayes Risk of α under Quasi prior for Data set II

β	γ	d	SELF		ELF		ABLF	
			$c = 0.5$	$c = 1.0$	$b_1 = 0.2$	$b_1 = 0.4$	$c_2 = 0.5$	$c_2 = -0.5$
1.0	1.5	0.3	0.00484	0.00969	0.80938	1.61877	0.00724	0.01307
		1.3	0.00468	0.00937	0.80950	1.61900	0.00689	0.01285
1.5	2.0	0.3	0.00245	0.00491	0.87729	1.75459	0.00310	0.00785
		1.3	0.00237	0.00475	0.87741	1.75483	0.00295	0.00772
2.0	2.5	0.3	0.00155	0.00311	0.92277	1.84554	0.00175	0.00558
		1.3	0.00151	0.00301	0.92289	1.84578	0.00167	0.00549
2.5	3.0	0.3	0.00110	0.00220	0.95754	1.91509	0.00113	0.00430
		1.3	0.00106	0.00213	0.95766	1.91532	0.00108	0.00423

SELF: squared error loss function, ELF: entropy loss function, and ABLF: Al-Bayyati's loss function.

Table 4: Bayes Risk of α under Extension of Jeffery's prior Data set II

β	γ	c_1	SELF		ELF		ABLF	
			$c = 0.5$	$c = 1.0$	$b_1 = 0.2$	$b_1 = 0.4$	$c_2 = 0.5$	$c_2 = -0.5$
1.0	1.5	0.4	0.00476	0.00953	0.80944	1.61888	0.00707	0.01296
		1.4	0.00445	0.00890	0.80969	1.61938	0.00638	0.01252
1.5	2.0	0.4	0.00241	0.00483	0.87735	1.75471	0.00302	0.00779
		1.4	0.00225	0.00451	0.87760	1.75521	0.00273	0.00752

2.0	2.5	0.4	0.00153	0.00306	0.92283	1.84566	0.00171	0.00553
		1.4	0.00143	0.00286	0.92308	1.84616	0.00154	0.00535
2.5	3.0	0.4	0.00108	0.00216	0.95760	1.91521	0.00111	0.00426
		1.4	0.00100	0.00202	0.95785	1.91570	0.00101	0.00412

SELF: squared error loss function, ELF: entropy loss function, and ABLF: Al-Bayyati's loss function.

From Table 3 and 4 shows that squared error loss function provides the minimum posterior risk as compared to the other loss functions particularly as C is (0.5) and the prior Extension of Jeffery's prior provides the less posterior risk than Quasi prior. Moreover, when we increase the true value of parameters $(\beta, \gamma) = c((1.0, 1.5), (1.5, 2.0)$ and $(2.5, 3.0))$ and increase the value of $d = c(0.3, 1.3)$ and $c_1 = c(0.4, 1.4)$, the Bayes risk of $\hat{\alpha}$ decreases quite significantly.

7. Conclusion:

On comparing the Bayes posterior risk of different loss functions, it is observed that SELF has less Bayes posterior risk than other loss functions in both priors. According to the decision rule of less Bayes posterior risk we conclude that SELF is more preferable loss function for different values of parameters.

It is clear from Tables 1& 4 the comparison of Bayes posterior risk under different loss functions using quasi as well as Extension of Jeffery's priors has been made through which we conclude that within each loss function Extension of Jeffery's prior provides less Bayes posterior risk than Quasi prior so it is more suitable for the generalized exponentiated moment exponential distribution. Moreover, when we increase the true value of parameters $(\beta, \gamma) = c((1.0, 1.5), (1.5, 2.0)$ and $(2.5, 3.0))$ and increase the value of $d = c(0.3, 1.3)$ and $c_1 = c(0.4, 1.4)$, the Bayes risk of $\hat{\alpha}$ decreases quite significantly.

References:

1. Al-Bayyti , H.N.(2002): "*Comparing methods of estimating Weibull failure models using simulation*", Ph.D. Thesis, College of Administration and Economics, Baghdad University, Iraq.

2. Al-Kutubi, H. S., (2005): "On comparison estimation procedures for parameter and survival function", *Iraqi Journal of Statistical Science*, vol. 9, 1-14.

978-1-62265-940-1 (online) 978-1-62265-941-8 (paper) - Applied Mathematics Research Cases

3. Calabria, R. and Pulcini, G., (1994): "An engineering approach to Bayes estimation for the Weibull distribution", *Microelectronics Reliability*, vol.34, no.5, pp.789-802.

4. Dara, S.T. and Ahmad, M., (2012): "Recent Advances in Moment Distributions and their Hazard Rate. Ph.D. Thesis", National College of Business Administration and Economics Lahore, Pakistan.

5. Devendera Kumar (2016): "Moments and Estimation of the Exponentiated Moment Exponential Distribution", Mathematical Sciences and Applications E-Notes, 4(1) 94-112.

6. Dey, D.K., Ghosh, M. and Srinivasan, C., (1987): "Simultaneous Estimation of Parameters under Entropy loss", *J. Statist.Plan. and Infer.*, pp. 347-363 .

7. Dey, D.K. and Liu, Pie-San L., (1992): "On Comparison of Estimators in a Generalized Life Model", *Micro-electron. Reliab.*, vol.45, no.3, pp. 207-221.

8. Gross, A.J. and Clark, V.A. (1975): "Survival Distributions: Reliability Applications in the Biometrical Sciences", John Wiley, New York.

9. Gupta, R.D., Kundu, D., (2009): "A new class of weighted exponential distribution. *Statistics"* **43**, 621-634.

10. Farahani, Z.S.M., Khorram, E., (2014): "Bayesian statistical inference for weighted exponential distribution", *Communication in Statistics-Simulation and Computation* **43**, 1362-1384.

11. Hasnain, S.A. (2013): "Exponentiated Moment Exponential Distribution. Ph.D. Thesis", National College of Business Administration and Economics, Lahore, Pakistan.

12. Meeker W. Q., Escobar L. A. (1998): "Statistical Methods for Reliability Data", *John Wiley, New York,* 383.

13. Sanku Dey, Sajid Ali and Chanseaok Park. (2015): "Weighted exponential distribution properties and different methods of estimation", *Journal of Statistical Computation and Simulation.*

14. Zafar Iqbal et *al* (2014): "Generalized Exponentiated Moment Exponential Distribution", Pak. J. Statist. Vol. 30(4), 537-554

978-1-62265-940-1 (online) 978-1-62265-941-8 (paper) - Applied Mathematics Research Cases

POLYGON DEFORMATION APPROACH - A NEW FORMULATION FOR THE TRAVELING SALESMAN PROBLEM

Elias Munapo

School of Economics and Decision Sciences, North West University, Mafikeng Campus, Mmabatho, 2735, South Africa. emunapo@gmail.com

ABSTRACT

The chapter presents a new approach for formulating the Traveling Salesman Problem (TSP) as a linear integer model. The approach assumes an optimal tour as a polygon that has been deformed. The outmost cities or nodes of the TSP are joined together to form a polygon. The polygon is then deformed by joining an interior node to any of the two outmost nodes. The process is repeated until all nodes inside the polygon are connected. Linear integer formulation is used to take care of all the various ways of identifying the interior nodes and connecting them to the outmost nodes. The proposed approach has the advantage that no sub-tours can form. Sub-tour elimination constraints that are usually used in other classes of TSP formulations are not necessary.

Keywords: NP hardness, traveling-salesman problem (TSP), linear integer model, polygon and deformation.

1 INTRODUCTION

The traveling salesman problem (TSP) is one of those well-known difficult problems. This is a problem of visiting each of the given *n* cities or nodes so that each city or node is visited only once, the total distance traveled is minimal and then return to original node base. This problem is old and was studied in the 18th century by an Irish mathematician, William Rowan Hamilton and a British mathematician named Thomas Penyngton Kirkman [2]. It is NP-hard and is still difficult to solve up to now. Heuristics such as the one by Basu [1] are used at the moment to come up with near optimal solutions for practical problems that occur in real life. The traveling salesman problem has many applications and that is the reason why it has received so much attention from a large number of researchers. The various areas of application of the TSP model include:

1.1. Routing problem

This class includes the rubbish collection, postal and delivery trucks. There is need for managers to know the routes that their trucks should follow so that distance, time and fuel use are minimized. As an example milk or bread has to be delivered to 40 shops in a single city by a single van. This type of a problem requires proper planning so that time and fuel are saved. Detailed information on this model can be obtained in Lenstra and Rinnooy Kan [10].

1.2. Order-picking problem in warehouses

It is not easy to pick items or orders from a very large store room or warehouse. If the items or goods are stored in such a way that each item has a fixed position then picking 5 or 6 items from the warehouse requires minimizing time and distance travelled. If nodes are used to represent the fixed positions in this problem then it becomes a TSP. For detailed discussions on this problem, see van Dal [15].

1.3. Servicing gas turbine engines

In gas turbine engines of an aircraft there are nozzle-guided vane assemblies located at each turbine stage. Since the vanes have individual characteristics then the problem of placing the vanes in the best way during servicing is a TSP. More on servicing gas turbine engines can be found in Plante et al. [13].

1.4. Drilling of circuit boards

Holes are required to connect a conductor on one layer of a printed circuit board to the other layer. If the holes are of different sizes or diameter then there is a need to change the size of the head of the drilling machine. Since it is time consuming to move to a tool box and change the head of a drilling machine then all holes of the same size or diameter have to be drilled before changing the head. The problem of drilling holes of the same size on the circuit board and minimizing distance moved can be formulated as a TSP [9].

1.5. X-ray crystallography

The TSP model is also used in analyzing the structure of crystals. This is done by using an X-ray diffractometer to obtain information about the structure of the crystalline material being analyzed. The problem of selecting the best position of the detector, motors and other gargets used in the analysis can be shown to be a TSP. For this analysis there is a very large number of possible positions to select from. Readers are encouraged to see Bland and Shallcross [3] for more information on crystallography.

The paper proposes an integer linear programming (LP) formulation of the TSP problem with equality constraints and nonnegative variables and with the property that any basic solution of the relaxed LP (without integer constraints) is integer. Since any LP can be solved in polynomial time, this settles (in the affirmative) one of the most important open problems in mathematics (and or computer science) of whether P=NP.

2. TSP AS A NETWORK

A TSP is very easy to explain as just a network model. The network model is presented in Figure1.

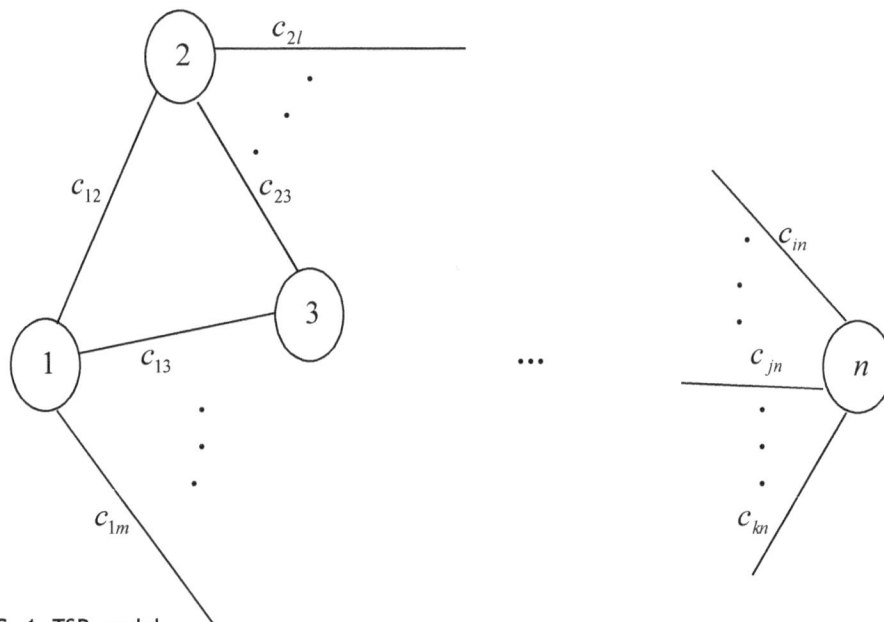

FIG. 1. TSP model

Where c_{ij} is the distance between node i and node j.

Let $N = \{1, 2, ..., n\}$ where $i, j, k, m \in N$.

The objective is to move from node 1 and back to node 1 in such a way that all nodes are visited and each node is visited only once.

$\forall \; i \neq j$, let

$$x_{ij} = \begin{cases} 1 & \text{if arc } (i,j) \text{ is a link in the tour} \\ 0 & \text{otherwise} \end{cases}$$

The TSP becomes very difficult when it is expressed as a mathematical model.

3. POLYGON AND POLYGON DEFORMATION

A polygon is formed by connecting only outermost nodes of a TSP by straight lines. Then polygon deformation can be defined as the process of connecting the polygon to those nodes that are inside the polygon.

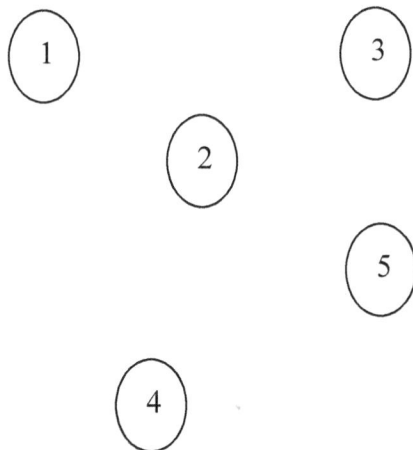

FIG. 2. Nodes of a TSP

The outermost nodes of the TSP can be connected to come up with a polygon as shown in Figure 3. The outmost arcs are also known as *boundary arcs*.

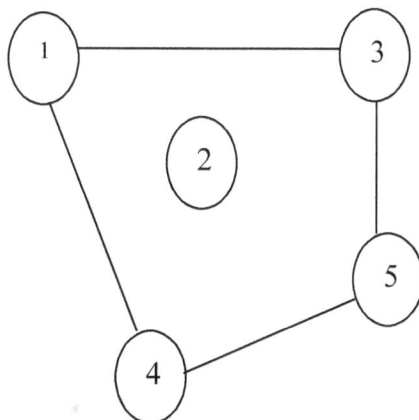

FIG 3. A polygon

The polygon is made up of outermost nodes and in this case node 2 is an interior node. For us to come up with a tour, then the interior node must be connected. There are 4 ways of doing it. Out of these ways, at least one will be the cheapest in terms of distance. In this paper the process of connecting the interior node to form a new polygon is called *polygon deformation*.

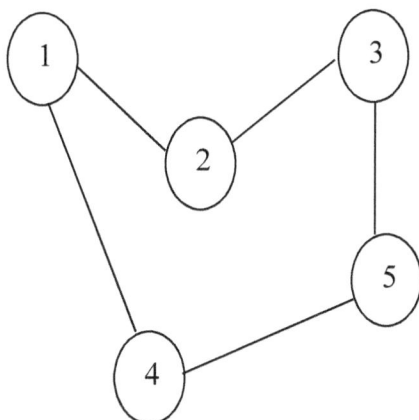

FIG. 4. Possible way 1 - when arc (1,3) is deformed.

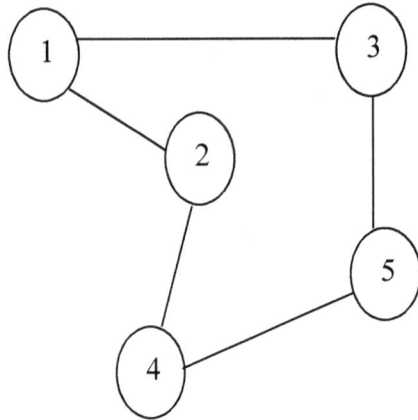

FIG. 5. Possible way 2 - when arc (1,4) is deformed.

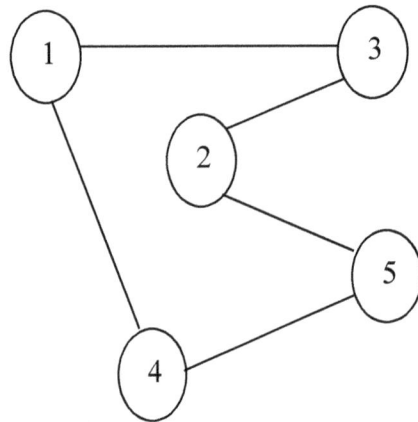

FIG. 6. Possible way 3 - when arc (3,5) is deformed.

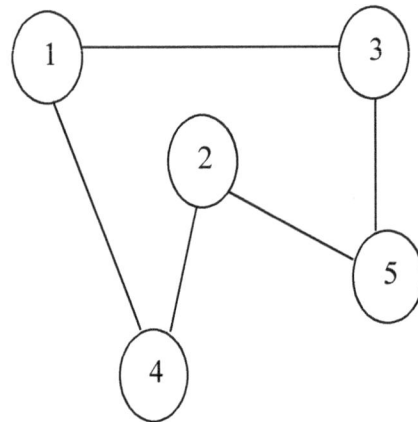

FIG. 7. Possible way 4 - when arc (4,5) is deformed.

At least one of these ways will give the cheapest in terms of distance but the problem is we do not know which one it is. This problem of identifying the correct deformation can be modeled as an integer model.

4. MODELING AS A LINEAR INTEGER MODEL

4.1. Arc deformation constraints

For practical problems, the number of ways of deforming an arc is not unique and is not straight forward and obvious. In this case binary variables are assigned to these ways and the sum of binary variables must be equal to 1 since only one way must be selected. These restrictions are called *arc deformation constraints*.

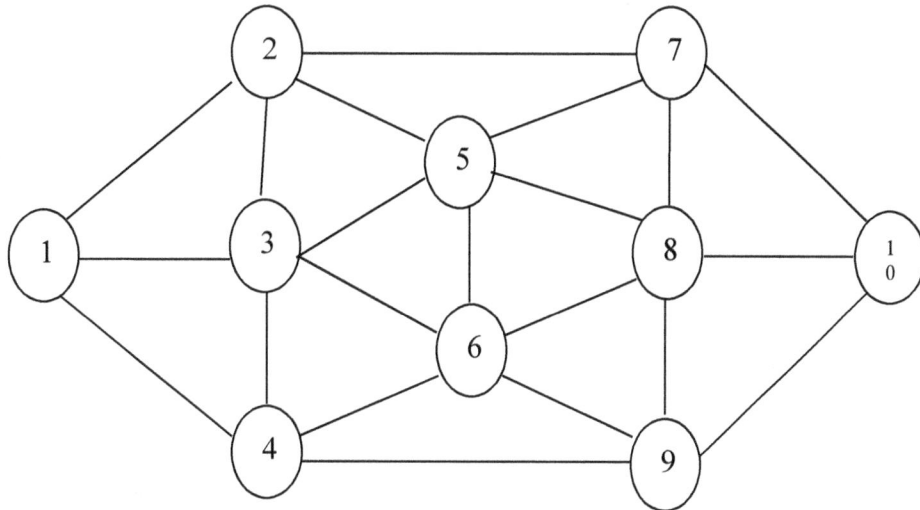

FIG. 8. An example of a TSP with 10 nodes

Let the distance of arc (i,j) in Figure 8, be c_{ij} and the binary variable representing this arc be x_{ij}. The polygon for this TSP network model is 1-2-7-10-9-4-1. Suppose the boundary arc (1,4) of this polygon is to be deformed. This arc can be deformed in 4 ways as shown in the following figures.

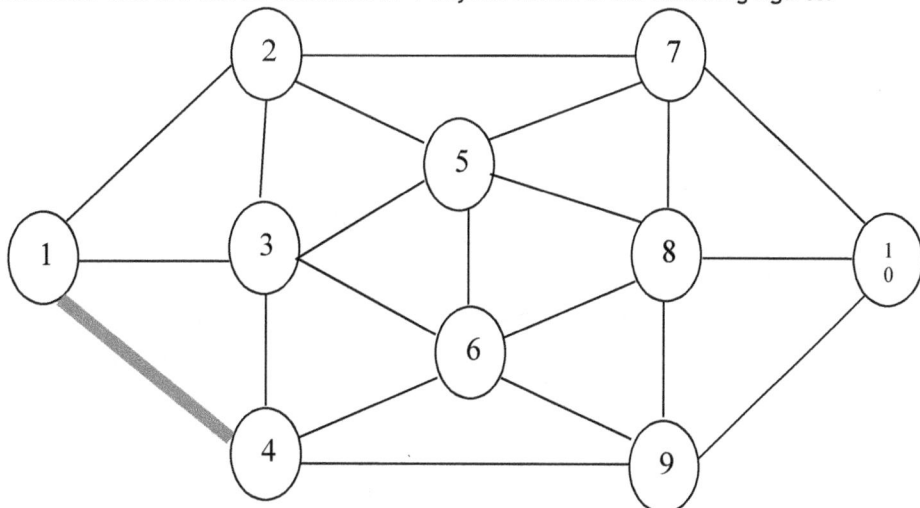

FIG. 9. Arc (1,4) deformation way 1 - when there is no deformation at all

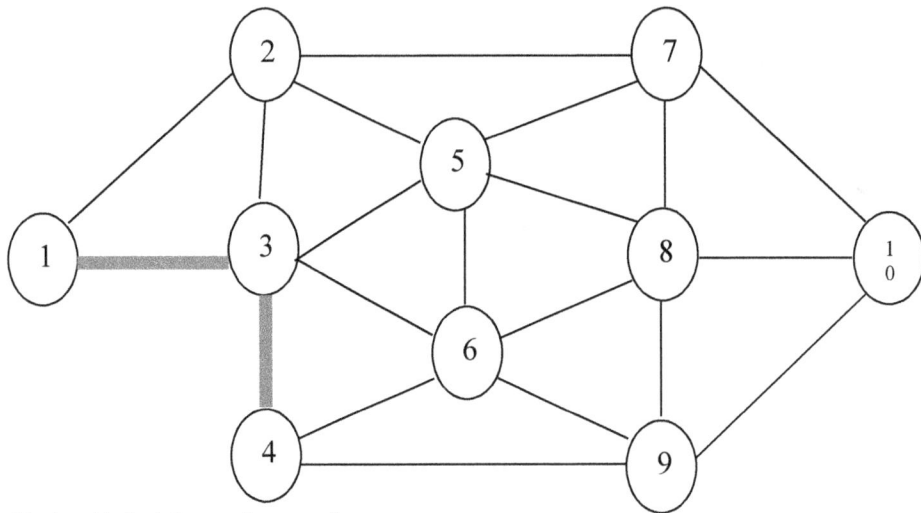

FIG. 10. Arc (1,4) deformation way 2

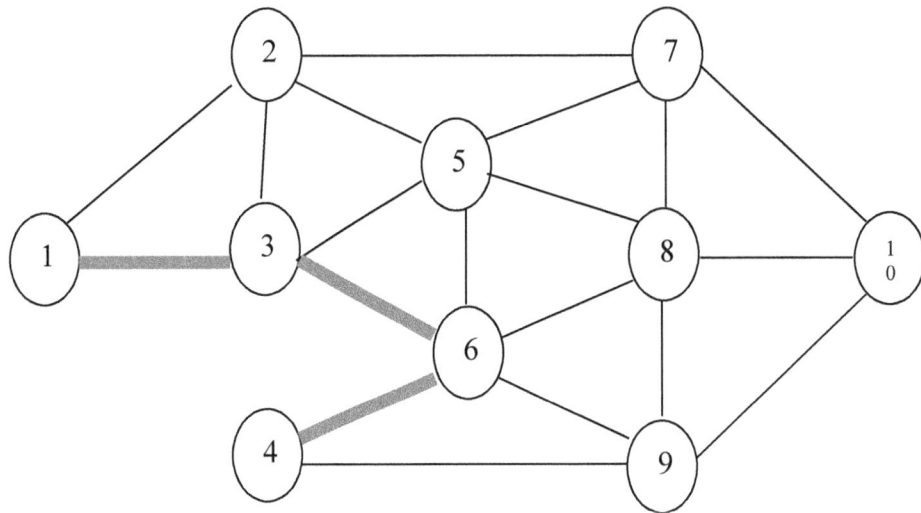

FIG. 11. Arc (1,4) deformation way 3

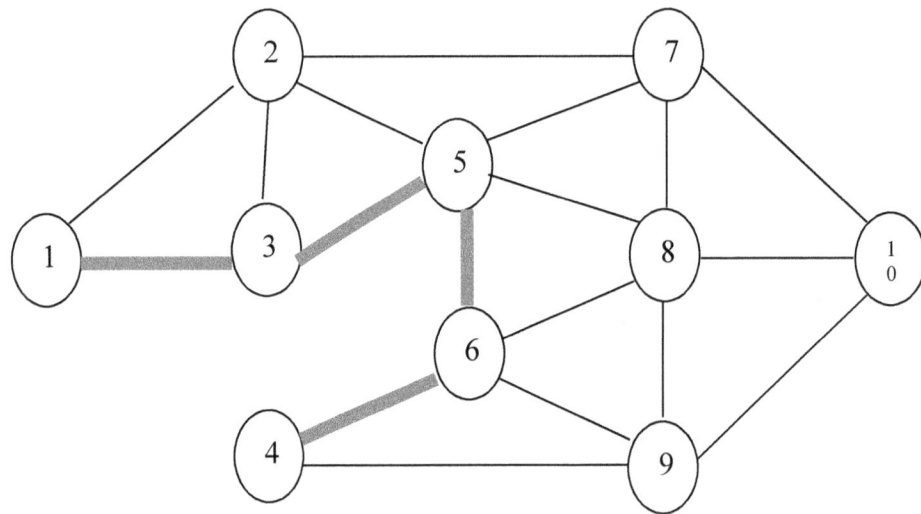

FIG. 12. Arc (1,4) deformation way 4

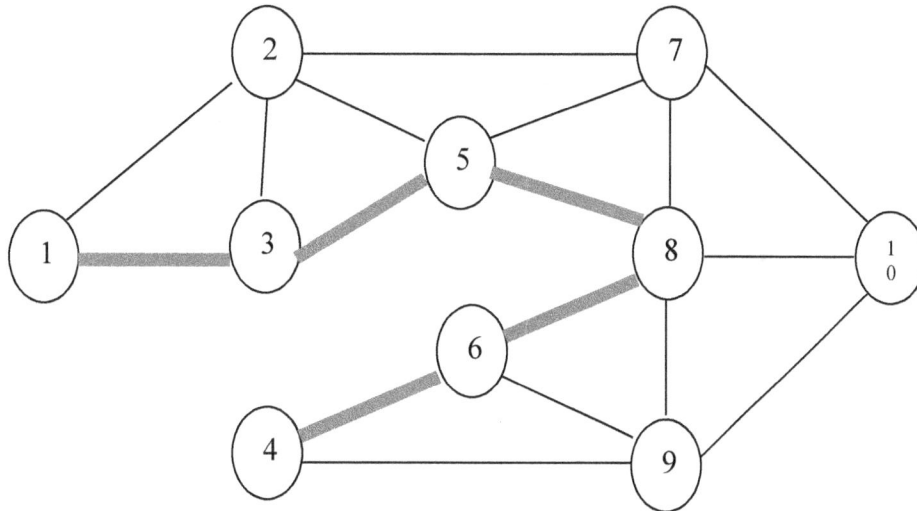

FIG. 13. Arc (1,4) deformation way 5

Let $\phi_k^{i,j}$ be the kth - way deformation binary variable for arc (i,j). The ways of deforming arc (1,4) can be represented in integer form as follows.

Way 1: $x_{14} \geq \phi_1^{1,4}$ This is when there is no arc deformation.

Way 2: $x_{13} + x_{34} \geq 2\phi_2^{1,4}$

Way 3: $x_{13} + x_{36} + x_{46} \geq 3\phi_3^{1,4}$

Way 4: $x_{13} + x_{35} + x_{56} + x_{64} \geq 4\phi_4^{1,4}$

Way 5: $x_{13} + x_{35} + x_{58} + x_{68} + x_{46} \geq 5\phi_5^{1,4}$

$$\phi_0^{1,4} + \phi_1^{1,4} + \phi_2^{1,4} + \phi_3^{1,4} + \phi_4^{1,4} + \phi_5^{1,4} = 1$$

For any boundary arc (i,j) in any TSP network model this can be generalized as

$$x_{ij} \geq \phi_0^{i,j}$$

$$...$$

$$x_{is} + ... + x_{tj} \geq k\phi_k^{i,j}$$

$$\phi_0^{i,j} + ... + \phi_k^{i,j} = 1$$

Where $s, t \in N$.

4.2. Openings and deformation depth

A boundary arc can be deformed if there is an *opening*. In other words an opening exist if it is possible to deform a given boundary arc. *Deformation depth* of an opening is the number of the arcs that can be crossed during the deformation process. These arcs include the boundary arc. The opening and deformation depth are illustrated in Figure 14.

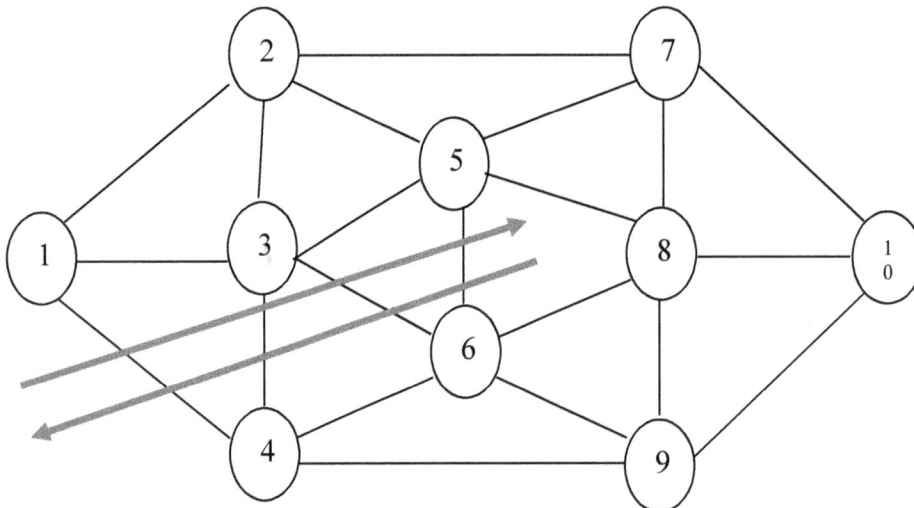

FIG. 14. Illustration of an opening and deformation depth.

For Figure 14, the boundary arc (1,4) has an opening and the opening is shown by two arrows. In this case the deformation depth = 4, meaning a maximum of 4 arcs can be crossed during the deformation process. Also note that in the boundary arc (1,4), has one opening, in some cases a boundary arc can have more than one openings.

4.3. Standard constraints

For us to have a tour, one arc must enter a node and another must leave that node. This can be formulated as a constraint. Such a constraint is called a *standard constraint*.

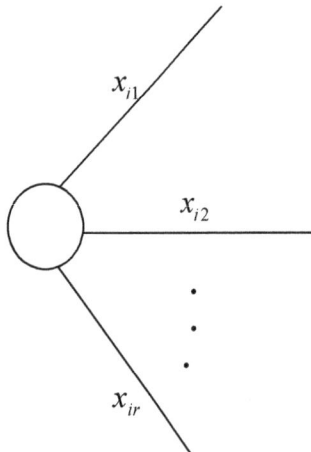

FIG. 15. Standard constraint

Suppose r is the number of arcs emanating from node i.

$$x_{i1} + x_{i2} + ... + x_{ir} = 2$$

Where $r \in N$.

These constraints alone cannot solve the TSP problem. If these constraints are used on their own then the optimal solution might have sub-tours.

5. THE POLYGON DEFORMATION LINEAR INTEGER MODEL FOR TSP MODEL

Given any TSP network model given in Figure 1, a polygon deformation linear integer model can be formulated as:

Minimise $\quad c_{12}x_{12} + c_{13}x_{13} + \ldots + c_{kn}x_{kn}$

Such that

$$\ldots$$

$$x_{i1} + x_{i2} + \ldots + x_{ir} = 2 \qquad \left.\right\} \text{Standard constraints}$$

$$\ldots$$

$$x_{ij} \geq \phi_0^{i,j}$$

$$\ldots$$

$$x_{is} + \ldots + x_{tj} \geq k\phi_k^{i,j} \qquad \left.\right\} \text{Arc deformation constraints}$$

$$\phi_0^{i,j} + \ldots + \phi_k^{i,j} = 1$$

$$\ldots$$

Where $\phi_k^{i,j}$ and x_{ij} are binary variables.

For this model no sub-tour elimination constraints are necessary.

5.1. Illustration

Formulate the polygon deformation linear integer model for the TSP network diagram given in Figure 16.

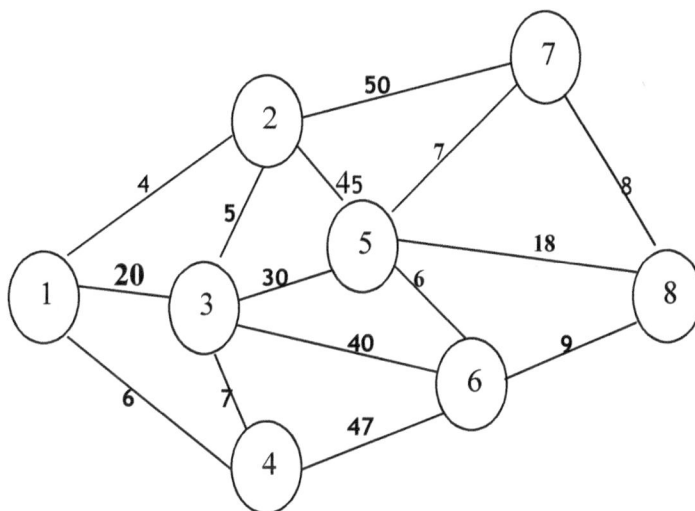

FIG. 16. Illustration

This illustration has two sub-tours that are possible if only the standard constraints are used in the formulation.

Sub-tour 1: 1-2-3-4-1
Sub-tour 2: 5-7-8-6-5

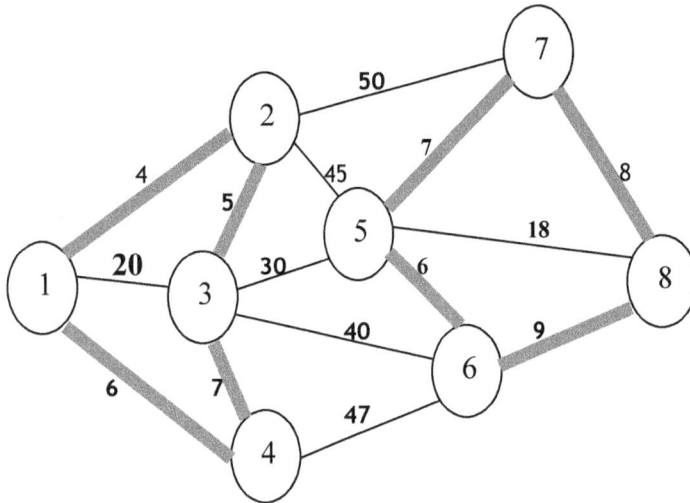

FIG. 17. Sub-tours

5.2. *The polygon deformation linear integer model*

Objective function: Minimize
$$4x_{12} + 20x_{13} + 6x_{14} + 5x_{23} + 45x_{25} + 50x_{27} + 7x_{34} + 30x_{35}$$
$$+ 40x_{36} + 47x_{46} + 6x_{56} + 7x_{57} + 18x_{58} + 9x_{68} + 8x_{78}$$

Standard constraints:

$x_{12} + x_{13} + x_{14} = 2$ (node 1)

$x_{12} + x_{23} + x_{25} + x_{27} = 2$ (node 2)

$x_{13} + x_{23} + x_{34} + x_{35} + x_{36} = 2$ (node 3)

$x_{14} + x_{34} + x_{46} = 2$ (node 4)

$x_{25} + x_{35} + x_{56} + x_{57} + x_{58} = 2$ (node 5)

$x_{36} + x_{46} + x_{56} + x_{68} = 2$ (node 6)

$x_{27} + x_{57} + x_{78} = 2$ (node 7)

$x_{58} + x_{68} + x_{78} = 2$ (node 8)

Arc deformation constraints:

$$\left. \begin{array}{l} x_{12} \geq \phi_1^{1,2} \\ x_{13} + x_{23} \geq 2\phi_2^{1,2} \\ x_{13} + x_{25} + x_{35} \geq 3\phi_3^{1,2} \\ \phi_1^{1,2} + \phi_2^{1,2} + \phi_3^{1,2} = 1 \end{array} \right\} \text{ boundary arc (1,2)}$$

$$\left. \begin{array}{l} x_{14} \geq \phi_1^{1,4} \\ x_{13} + x_{34} \geq 2\phi_2^{1,4} \\ \phi_1^{1,4} + \phi_2^{1,4} = 1 \end{array} \right\} \text{ boundary arc (1,4)}$$

25

$$x_{27} \geq \phi_1^{2,7}$$
$$x_{25} + x_{57} \geq 2\phi_2^{2,7}$$
$$x_{23} + x_{35} + x_{57} \geq 3\phi_3^{2,7}$$
$$\phi_1^{2,7} + \phi_2^{2,7} + \phi_3^{2,7} = 1$$

$\left.\right\}$ boundary arc (2,7)

$$x_{46} \geq \phi_1^{4,6}$$
$$x_{34} + x_{36} \geq 2\phi_2^{3,6}$$
$$x_{34} + x_{35} + x_{56} \geq 3\phi_3^{4,6}$$
$$\phi_1^{4,6} + \phi_2^{4,6} + \phi_3^{4,6} = 1$$

$\left.\right\}$ boundary arc (4,6)

$$x_{68} \geq \phi_1^{6,8}$$
$$x_{56} + x_{58} \geq 2\phi_2^{6,8}$$
$$x_{35} + x_{36} + x_{58} \geq 3\phi_3^{6,8}$$
$$\phi_1^{6,8} + \phi_2^{6,8} + \phi_3^{6,8} = 1$$

$\left.\right\}$ boundary arc (6,8)

$$x_{78} \geq \phi_1^{7,8}$$
$$x_{57} + x_{58} \geq 2\phi_2^{7,8}$$
$$\phi_1^{7,8} + \phi_2^{7,8} = 1$$

$\left.\right\}$ boundary arc (7,8)

Binary variables

$\phi_k^{i,j}$ and x_{ij} are binary variables.

5.3. Solution

Solving the binary linear integer model the optimal solution is obtained as:

$$x_{12} = x_{14} = x_{23} = x_{35} = x_{46} = x_{57} = x_{68} = x_{78} = 1$$
$$x_{13} = x_{23} = x_{25} = x_{27} = x_{34} = x_{36} = x_{56} = x_{58} = 0$$

Minimal distance $= 4 + 5 + 30 + 7 + 8 + 9 + 47 + 6 = 116.$

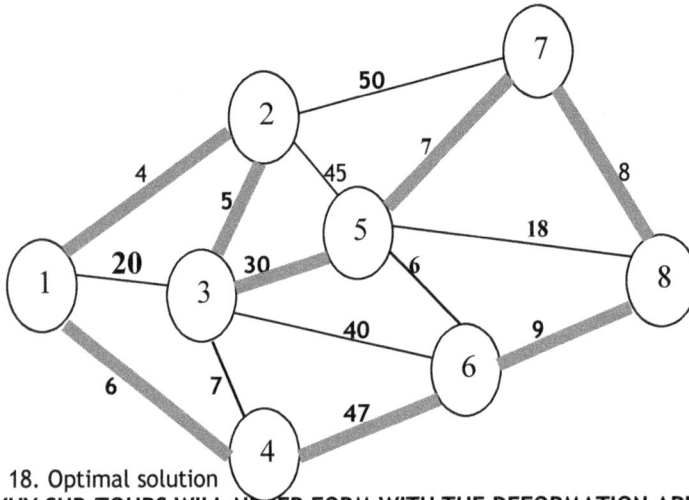

FIG. 18. Optimal solution

6. WHY SUB-TOURS WILL NEVER FORM WITH THE DEFORMATION APPROACH

6.1. Joining the outmost nodes forms a polygon. This polygon does not contain sub-tours.

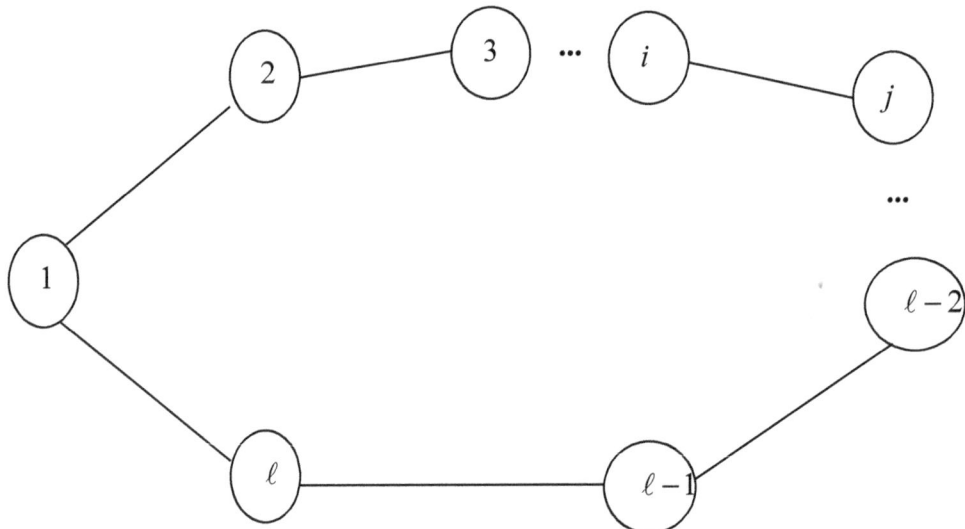

FIG. 19. Polygon formation by joining the outmost nodes of a TSP

6.2 Deforming an arc (i,j) into $i - i_1 - i_2 - i_3 - j$ does not form a sub-tour

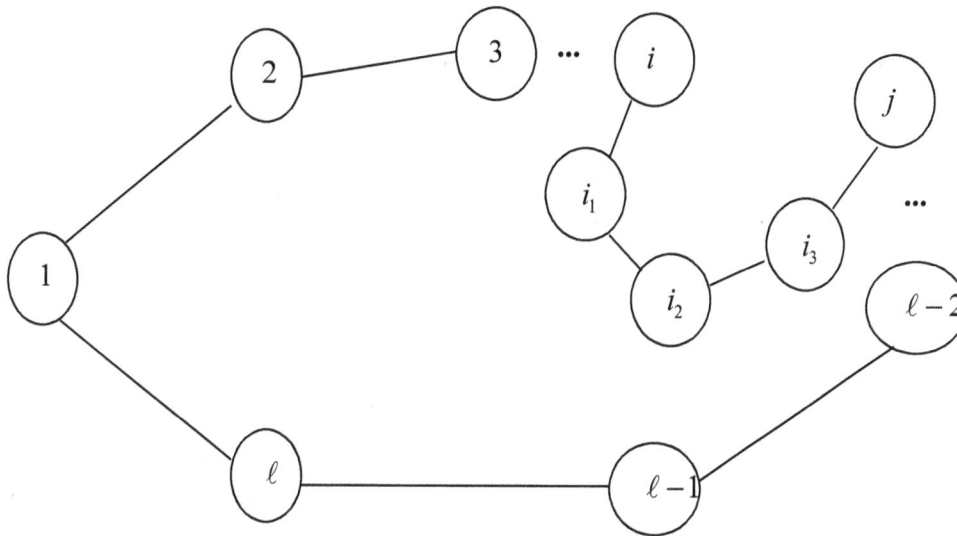

FIG. 20. Deforming arc (i,j) does not result in a sub-tour

7. OTHER LINEAR INTEGER FORMULATIONS FOR TSP

There are several linear integer formulations for the TSP model available in literature. These formulations can be classified as conventional, sequential, flow based and time staged. We summarize these formulations in this section and readers who may want more on these formulations can be referred to Orman and Williams [12].

7.1. Conventional formulation

This formulation was proposed by Dantzig et al. in 1954 [5]. The linear integer model is represented as:

Minimize $c_{12}x_{12} + c_{13}x_{13} + ... + c_{kn}x_{kn}$

$$\sum_{j} x_{ij} = 1$$
$$\sum_{i} x_{ij} = 1$$
Are called assignment constraints.

$$\sum_{i,j \in S} x_{ij} \leq |S| - 1, |S| \geq 2, S \subseteq N \setminus \{1\}, S \neq \{1\}$$ Are called sub-tour elimination constraints.

$$i \neq j.$$

The weakness of this approach is the large number of constraints. As n increases, the number of constraints increases exponentially.

7.2. Sequential formulation

In the sequential formulation the objective function and assignment constraints are the same as in the conventional, the only difference is in the sub-tour elimination constraints where continuous variables are used.

Minimize $c_{12}x_{12} + c_{13}x_{13} + ... + c_{kn}x_{kn}$

$$\sum_j x_{ij} = 1$$

$$\sum_i x_{ij} = 1$$

$u_i - u_j + nx_{ij} \le n-1, \ \forall ij \in N \setminus \{1\}.$ Sub-tour elimination constraints.

$u_i, u_j \ge 0$ are continuous variables.

$i \ne j.$

Unlike in the previous case the increase in the number of sub-tour elimination constraints for this formulation is polynomial. See Miller et al. [11] for more information.

7.3. Flow based formulations

The flow based formulations have variants or versions. In this class we may have *a single commodity, two commodity* or *multi-commodity* flow formulations. The formulations under this category like the sequential formulation also use continuous variables like in the sequential formulation. The flow based formulations use flow constraints as sub-tour constraints. The assignment constraints are also retained as in the previous case. The three types of flow based formulations differ in the nature of flow constraints. More on these flow constraints can be obtained from Claus [4], Gavish and Graves [8], Finke et al. [6] or Wong [16]. Only the single commodity flow is used to explain the flow based formulation in this paper.

Let y_{ij} = flow in arc (i,j). Then single commodity flow formulation can be represented can be represented as:

Minimize $c_{12}x_{12} + c_{13}x_{13} + ... + c_{kn}x_{kn}$

$$\sum_j x_{ij} = 1$$

$$\sum_i x_{ij} = 1$$

$$y_{ij} \le (n-1)x_{ij}$$

$$\sum_j y_{1j} = n-1$$

$$\sum_i y_{ij} - \sum_l y_{jl} = 1 \qquad \forall j \in N \setminus \{1\}, i \ne l.$$

Are called flow constraints.

$i \ne j.$

The number of constraints is also better than that of the conventional formulation and is also polynomial.

7.4. Time staged formulations

In this formulation a new binary variable (y_{ij}^t) is introduced to represent a stage (t) in which an arc (i,j) is traversed. In other words the binary variable (y_{ij}^t) will assume values as shown below and satisfying the following time staged constraints. The assignment constraints do not change. There are several variants of this formulation. We can have either 1st stage, 2nd stage or a higher stage of formulation and readers are encouraged to see Fox et al. [7] or Vajda [14] for detailed discussions. In this paper we concentrate mainly on stage 1 to explain this class of formulation.

$$y_{ij} = \begin{cases} 1 & \text{if arc(i.j) is traversed at stage t} \\ 0 & \text{otherwise} \end{cases}$$

Minimize $\quad c_{12}x_{12} + c_{13}x_{13} + \ldots + c_{kn}x_{kn}$

$$\sum_j x_{ij} = 1$$

$$\sum_i x_{ij} = 1$$

$$\sum_{i,j,t} y_{ij}^t = n$$

$$\sum_{j, i \geq 2} t y_{ij}^t - \sum_{k,t} t y_{ki}^t = 1 \qquad \forall\, i \in N \setminus \{1\}$$

$$x_{ij} - \sum_t y_{ij}^t = 0$$

Are called time staged constraints.

$$i \neq j.$$

This class of formulation has also a polynomial number of constraints.

8. CONCLUSIONS

The TSP model is difficult to solve and because of its many important applications it has attracted the attention of many researchers. The proposed formulation presented in this chapter is a promising formulation for solving the TSP model. The formulated integer linear programming model can be solved with no possibilities of sub-tours. Sub-tour elimination constraints that are usually used in the other TSP formulations are not necessary. The proposed approach provides an extra avenue for the hunt of an efficient algorithm for the difficult TSP model.

REFERENCES

[1] S. Basu, Tabu Search Implementation on Traveling Salesman Problem and Its Variations: A Literature Survey, American Journal of Operations Research 2(2) (2012), 163-173.

[2] N.L. Biggs, K.E. Lloyd and R.J. Wilson, Graph Theory Clarendon Press, Oxford, 1986, pp. 1736-1936.

[3] R.E. Bland and D.E. Shallcross, Large traveling salesman problem arising from experiments in X-ray crystallography, a preliminary report on computation, Operations Research Letters 8(3) (1989), 125-128.

[4] A. Claus, A new formulation for the travelling salesman problem, SIAM J. Alg. Disc. Math. 5 (1984), 21-25.

[5] G.B. Dantzig, D.R. Fulkerson and S.M. Johnson, Solutions of a large scale travelling salesman Problem, Ops. Res. 2 (1954), 393-410.

[6] G. Finke, A. Claus and E. Gunn, A two-commodity network flow approach to the travelling salesman problem, Combinatorics, Graph Theory and Computing, Proc.14th South Eastern Conf., Atlantic University, Florida, 1983.

[7] K.R. Fox, B. Gavish and S.C. Graves, An n-constraint formulation of the (timedependent) travelling salesman problem, Ops. Res. 28 (1980), 1018-1021.

[8] B. Gavish and S.C. Graves, The traveling salesman problem and related problems, Working Paper OR-078-78, Operations Research Center, MIT, Cambridge MA, 1978.

[9] M. Grötschel, M. Jünger and G. Reinelt, Optimal Control of Plotting and Drilling Machines: A Case Study, Mathematical Methods of Operations Research 35(1) (1991), 61-84.

[10] J.K. Lenstra and A.H.G Rinnooy Kan, Some simple applications of the traveling salesman Problem, Operational Research Quarterly 26 (1975), 717-33.

[11] C.E. Miller, A.W. Tucker and R.A. Zemlin, Integer programming formulation of travelling salesman problems, J. ACM 3 (1960), 326-329.

[12] A.J. Orman and H.P. Williams, A survey of different integer programming formulations of the travelling salesman problem. In: Kontoghiorghes E. & Gatu C. (eds). Optimisation, Econometric and Financial Analysis Advances in Computational Management Science, Springer: Berlin, Heidelberg, 2006, pp. 91-104.

[13] R.D. Plante, T.J. Lowe and R. Chandrasekaran, The Product Matrix Traveling Salesman Problem: An Application and Solution Heuristics, Operations Research 35 (1987), 772-783.

[14] S. Vajda, Mathematical Programming, Addison-Wesley, London, 1961.

[15] R. van Dal, Special Cases of the Traveling Salesman Problem. Wolters-Noordhoff, Groningen, 1992.

[16] R.T. Wong, Integer programming formulations of the travelling salesman problem, Proc. IEEE Conf. on Circuits and Computers, 1980, pp. 149-152.

978-1-62265-940-1 (online) 978-1-62265-941-8 (paper) - Applied Mathematics Research Cases

Stability Analysis of Predatorprey model with ratio-dependent functional response

Ahmed Buseri Ashine

Department of Mathematics, MaddaWalabu University, Bale Robe - Ethiopia

Abstract

This paper concerns with a two dimensional nonlinear dynamical predator-prey model with ratio-dependent functional response. Dynamical analysis involving determination of equilibrium points on their local stabilities is presented.

Keywords: predator-prey, ratio-dependent, equilibrium points, local stability

1. Introduction

Predator-prey behavior is a form of very common biological interaction in nature. Mathematical model for predator-prey interaction is studied originally by Lotka [2] and Volterra [6] and is known as Lotka-Volterra model. The model is only consider four factors such as growth rate of prey, predation rate, mortality rate of predator, and conversion rate to change prey biomass into predator reproduction. Notice that all of the rates are linear. However, in the real life, predator-prey interaction does not depend only on those factors. Therefore, much developments of the model are proposed based on biological assumptions in the real life.

According to some biologists, such as Arditi and Ginzburg [4], ecological functional response should depend on the density of prey and predator, since predators occasionally have to search and compete for the prey. One of the functional responses which depend on the density of prey and predator is ratio-dependent functional response (see Xiao and Ruan [3], Edwin [1]). Therefore, in this paper we concern with dynamical analysis of predator-prey model with ratio-dependent response function. It is assumed that prey as well as predator grows logistically, since predator has other food source besides prey. Hence, the predator has two growth rate, namely logistic and predation growth. In order to control the amount of predator population, it is assumed that a linear rate of harvesting is applied to predator population.

2. The Model

Predator-prey model in this paper modifies the model discussed by Kar and Chaudhuri [5] by replacing Holling type II functional response by ratio-dependent functional response. Hence, the model that we concern with is

$$\begin{cases} \dfrac{dX}{dt} = r\left(1-\dfrac{X}{K_1}\right)X - \dfrac{aXY}{Y+bX} \\[2ex] \dfrac{dY}{dt} = s\left(1-\dfrac{Y}{K_2}\right)Y + \dfrac{cXY}{Y+bX} \end{cases} \tag{1}$$

Where X represents prey density, Y is predator density, r and s are growth rate of prey and predator respectively, K_1 and K_2 represent carrying capacity of prey and predator respectively, a is parameter of capturing rate predator on prey, $1/b$ is Michaelis-Menten constant, and c represents conversion rate to change prey biomass into predator reproduction.

3. Equilibria

The possible equilibrium points of system (1) are $E_1\,(0, K_2)$, $E_2\,(K_1, 0)$, and $E_3\,(X^*, Y^*)$ where

$$X^* = \frac{-B+\sqrt{B^2-4AC}}{2A} \text{ or } X^* = \frac{-B-\sqrt{B^2-4AC}}{2A}, \text{ and}$$

$$Y^* = \frac{rbX^*\left(1-\dfrac{X^*}{K_1}\right)}{a-r\left(1-\dfrac{X^*}{K_1}\right)}.$$

Here,
$$A = \frac{r}{K_1}\left(\frac{sb}{K_2}+\frac{cr}{abK_1}\right),$$

$$B = \frac{r}{K_1}\left(s+\frac{2c}{ab}(a-r)\right)-\frac{rsb}{K_2},$$

$$C = (a-r)\left(s + \frac{c}{ab}(a-r)\right), \text{ and}$$

$$D = \left(\frac{rs}{K_1} - \frac{rsb}{K_2}\right) - \frac{rrsb}{K_1 K_2}(a-r)\left(s + \frac{c}{b}\right)$$

Proposition 1: Equilibrium point $E_3(X^*,Y^*)$ exists if one of the following conditions satisfied.

$$r > a, \tag{2}$$

or

$$r < a, \text{ B< 0, and D>0}, \tag{3}$$

or

$$r = a, \text{ and } 1 < \frac{K_1 b}{K_2} \tag{4}$$

Remark: X^* and Y^* satisfies the following equations.

$$r\left(1 - \frac{X^*}{K_1}\right) = \frac{aY^*}{Y^* + bX^*} \tag{5}$$

and

$$r\left(1 - \frac{Y^*}{K_2}\right) = -\frac{cX^*}{Y^* + bX^*} \tag{6}$$

4. Local Stability Analysis

Stability of equilibrium points is investigated by doing linearization on the system (1) around each equilibrium points.

Theorem 4.1: The equilibrium point E_1 (0, K_2) is locally asymptotically stable if $r < a$.

Proof: At E_1 (0, K_2), the Jacobean matrix becomes

$$J(E_1) = \begin{pmatrix} r-a & 0 \\ c & -s \end{pmatrix}$$

Jacobian matrix of E_1 has negativeEigen value for $r < a$. HenceE_1 (0, K_2) is locally asymptotically stable if $r < a$ and unstable if$r > a$.

Theorem 4.2: The equilibrium point E_2 (K_1, 0) is unstable.

Proof: At E_2 (K_1, 0), the Jacobean matrix is

$$J(E_2) = \begin{pmatrix} -r & -\dfrac{a}{b} \\ 0 & s+\dfrac{c}{b} \end{pmatrix}$$

Since $s + \dfrac{c}{b} > 0$, equilibrium point $E_2$is unstable.

Theorem 4.3: The equilibrium point E_3 (X^*, Y^*) is locally asymptotically stable provided $\Delta > 0$ and $\Gamma < 0$ Where $\Delta = xw - yz$ and $\Gamma = x + w$.

Proof: AtE_3 (X^*, Y^*), the Jacobean matrix becomes

$$J(E_3) = \begin{bmatrix} x & y \\ z & w \end{bmatrix}$$

Where

$$x = r\left(1 - \frac{2X^*}{K_1}\right) - a\left(\frac{Y^*}{Y^* + bX^*}\right)^2,$$

$$y = -ab\left(\frac{X^*}{Y^* + bX^*}\right)^2,$$

$$z = c\left(\frac{Y^*}{Y^* + bX^*}\right)^2,$$

$$w = s\left(1 - \frac{2Y^*}{K_2}\right) - bc\left(\frac{X^*}{Y^* + bX^*}\right)^2.$$

It is clear that $y < 0$ and $z > 0$. By substituting (6) into w, it is readily seen that $w < 0$. Under condition

(2) and (4), by substituting (5) into x, we get $x < -r\left(\dfrac{N^*}{K_1}\right)^2 < 0$ whereas under condition (3), x can be

negative or positive. For condition (2) and (4), theorem 4.3 is fulfilled because x, y, and w are negatives and $z > 0$. Hence, E_3 (X^*, Y^*) islocally asymptotically stable if condition (2) or (4) satisfied.

5. Conclusion

The model of predator-prey ratio-dependent response function is a system of two-dimensional nonlinear ordinary differentialequations. The system has three equilibrium point, namely the prey extinction point E_1 (0, K_2), the predator extinction pointE_2 (K_1, 0), and the survival pointE_3 (X^*, Y^*). Based on the analysis, E_2 (K_1, 0) is unstable. While, E_1 (0, K_2) and E_3 (X^*, Y^*) are local asymptotically stablewith certain conditions.

References

[1]   A. Edwin, Modeling and Analysis of a Two Prey - One Predator Systemwith Harvesting, Holling Type II and Ratio-dependent Responses, Master of Science in Mathematics of Makerere University, Uganda, 2010.

[2]   A.J. Lotka, Elements of Physical Biology, Williams and Wilkins Company, Baltimore, 1925.

[3]   G. Xiao and S. Ruan, Global Dynamics of a Ratio-dependent Predator-preySystem,*J. Math. Biol.*, 43 (2001), 268–290.

[4]   R. Arditi and L. R. Ginzburg, Coupling in Predator-prey Dynamics: Ratio-dependence, J. Theoret. Biol., 139 (1989), 311-326.

[5]   T.K. Kar and K.S. Chaudhuri, On Non-selective Harvesting of aMultispecies Fishery, International Journal of Mathematical Education inScience and Technology, 33 (2002), 543-556.

[6]   V. Volterra, Variations and Fluctuations in The Numbers of CoexistingAnimal Species, Lecture notes in Biomathematics, 1927, 65-273.

978-1-62265-940-1 (online) 978-1-62265-941-8 (paper) - Applied Mathematics Research Cases

Predator-Prey Interactions With Desease In Predator Incorporating Harvesting Of Predator

Ahmed Buseri Ashine

Department of Mathematics, MaddaWalabu University, Bale Robe – Ethiopia

Abstract: In this article, predator prey interactions where the predator is exposed to the risk of disease and harvesting is proposed. Equilibrium points, boundedness, and non-periodic solutions of the modelare obtained.Local stability and global stability were discussed. The equilibrium was stable locally, but not globally.

Keywords: prey-predator, stability, harvesting, Dulac's criterion

1. Introduction

Prey-predator models are of great interest to researchers in mathematics and ecology because they deal with environmental problems such as community's morbidity and how to control it, optimal harvest policy to sustain a community, and others. In the physical sciences, generic models can be constructed to explain a variety of phenomena. However, in the life sciences a model only describes a particular situation. Simple models such as the Lotka-Volterra are not able to tell us what is going on in the majority of cases. One of the reasons is due to the complexity of the biological ecosystem. Hence,we still seek for a variety of models to describe nature.

Theoretical and numerical studies of these models are able to give us an understanding of the interactions that is taking place. A particular class of models considers the existence of a disease in the predator or prey. Several models were constructed to study particular cases. To ensure the existence of the species involved, one of the steps taken is to harvest the infected species. In this paper, we consider the case where the infected predator is harvested. Several related theoretical studies have been conducted.

Amongst them are studies on the disease spread among the prey and the epidemic among predators with action incidence [6], the role of transmissible disease in the Holling Tanner predator prey model [4], the analysis of prey predator model with disease in the prey [7], another's study the disease in LotkaVolterra, [3]study the dynamics of a fisher resource system in an aquatic environment in two zones harvest in reserve area,[5] study the harvesting of infected prey,[1]show the stability analysis of harvesting,[2] Study the stability of harvested when the disease affects the predator by using the reproduction number.

The model is introduced after this section, followed by analysis onboundedness and properties of the solutions.

2. The Mathematical model

Consider the following dynamical system:

$$\begin{cases} \dfrac{dx}{dt} = a(1-x)x - b(y+z)x \\[2mm] \dfrac{dy}{dt} = cxy + \alpha yz - h_1 y \\[2mm] \dfrac{dz}{dt} = cxz - \alpha yz - h_2 z \end{cases} \tag{1}$$

where x, y, z are the prey, infected predator and susceptible predator respectively; a is the growth rate of prey; b, c the capture rate $(b > c)$, is the contact rate between the susceptible and infected predator; h_1, and h_2 are the harvest rates of the infected and susceptible predator respectively, and assume that the less effective predator shall be easier to harvest, so; it is better also to assume infected predator not become susceptible again and finally the disease does not affect the ability of the infected predator attacking prey.

2.1 Bounded of solutions

Theorem 1. The solution of system (1) is bounded.

Proof

Let the function $w(x,y,z) = x(t) + y(t) + z(t)$ and let μ be a positive number such that $0 < \eta < h_2$.

Then, $w'(t) + uw = ax(1-x) + \eta x - (b-c)(y+z)x - (h_1 - \eta)y - (h_2 - \eta)z$

$$w'(t) + uw < -a\left(x^2 - \left(\frac{a+\eta}{a}\right)x + \left(\frac{a+\eta}{2a}\right)^2 \right) + \frac{1}{a}\left(\frac{a+\eta}{2}\right)^2$$

$$\text{Let } \frac{1}{a}\left(\frac{a+\eta}{2}\right)^2 = v$$

$$w'(t) + uw(t) \le v$$

$$0 < w(x,y,z) \le \frac{v}{u}\left(1 - e^{ut}\right)(x,y,z)\big|_{t\to 0}$$

Theorem 2. Let $F(x, y) = ax - ax^2 - bxy$, $G(x, y) = cxy - h_1 y$,

$$M(x, z) = ax - ax^2 - bxz, \quad N(x, z) = cxz - h_2 z.$$

Define a function $H(x, y) = \dfrac{1}{xz}$.

Then $Q(x, y) = \dfrac{\partial(HF)}{\partial x} + \dfrac{\partial(HG)}{\partial y} = -\dfrac{a}{y}$

It's clear that is no change in change sign; therefore, this system cannot have any periodic solution in the xy-plane.

Again, $Q(x, z) = \dfrac{\partial(HM)}{\partial x} + \dfrac{\partial(HN)}{\partial z} = -\dfrac{a}{z}$

There is no change in sign; therefore, there is no periodic solution in xz- plane.

Hence, the system has no periodic solution.

2.2 Equilibrium

The dynamical system (1) has the following five fixed points: the origin(E_1), a predator fee fixed point (E_2), a disease free fixed point (E_3), a fixed point when all predator infected (E_4), and a fixed for which both population survive(E_5):

$$E_1 : (x, y, z) = (0,0,0)$$

$$E_2 : (x, y, z) = (1,0,0)$$

$$E_3 : (x, y, z) = (x_2, 0, z_2); \text{ where } x_2 = \frac{h_2}{c}, z_2 = \frac{a(1-x)}{b}$$

$$E_4 : (x, y, z) = (x_3, y_3, 0); \text{ where } x_3 = \frac{h_1}{c}, y_3 = \frac{a(1-x)}{b}$$

$$E_5 : (x^*, y^*, z^*) = \left(1 - \frac{b}{a\alpha}(h_1 - h_2), \frac{cx^* - h_2}{\alpha}, \frac{h_1 - bx^*}{\alpha}\right)$$

3. Stability

The Jacobian matrix of system (1) is given by:

$$J(x,y,z) = \begin{pmatrix} a-2ax-b(y+z) & -bx & -bx \\ cy & cx+\alpha z-h_1 & \alpha y \\ cz & -\alpha z & cx-\alpha y-h_2 \end{pmatrix}$$

Case 1. The system without Disease

When infected predators eradicate, the system (1) becomes:

$$\begin{cases} \dfrac{dx}{dt} = a(1-x)x - bxz \\ \dfrac{dz}{dt} = cxz - h_2 z \end{cases} \tag{1a}$$

The equilibrium (nontrivial) are $E_c'(1,0)$, $E_c'(x_2,z_2)$ where, $x_2 = \dfrac{h_2}{c}, z_2 = \dfrac{a}{b}(1-x_2)$

Proposition 1. $E_c'(1,0)$ is stable when $h_2 > c$ and unstable otherwise.

Proof: The eigenvalues near the first equilibrium are $-a$ and $c-h_2$. This completes the proof.

Theorem 3. If the equilibrium $E_c'(x_2,z_2)$ is locally stable, then the basin of attraction of this

equilibrium is denoted by $B(E_c'(x_2,z_2))$,

where $B(E_2') = \{(x,z) \in \Re_+^2 : x > \dfrac{h_2}{c}, z > \dfrac{a}{b}(1-x)with\dfrac{h_2}{c}z < \dfrac{a}{b}(1-x)x\}$

Proof: Let $V(x,z)$ be a function where

$$V(x,z) = \left(x-x_1 - x_2 \log^{\frac{x}{x_2}} \right) + \left(z-z_2 - z_2 \log^{\frac{z}{z_2}} \right), \text{ the}$$

$$\frac{dV}{dt} = -a(x-x_2)^2 - (b-c)(x-x_2)(z-z_2) < 0$$

The eigenvalues near $E_2'(x_2,z_2)$ are $\dfrac{-ax_2}{2} \pm \dfrac{\sqrt{a^2 x_2^2 - 4ah_2(1-x_2)}}{2}$ and $h_2 + \alpha z_2 - h_1$, and

stable when $1 - \dfrac{a\alpha(1-x_2)}{b(h_1-h_2)} > 0$

Case 2. When all predators become infected

When all predators become infected the subsystem of system (1) becomes:

$$\begin{cases} \dfrac{dx}{dt} = a(1-x)x - bxy \\ \dfrac{dy}{dt} = cxy - h_1 y \end{cases} \qquad (1b)$$

The equilibrium (nontrivial) are $E_c{}'(1,0)$, $E_c{}'(x_3,y_3)$ where, $x_3 = \dfrac{h_1}{c}$, $y_3 = \dfrac{a}{b}(1-x_3)$

Proposition 2. $E_c{}'(1,0)$ is stable when $h_1 > c$ and unstable otherwise.

Proof: The eigenvalues near $E_c{}'(1,0)$ are $-a$ and $c - h_1$. This completes the proof.

Theorem 4. Assume the equilibrium $E_c{}'(x_3,y_3)$ is locally stable, the basin of attraction of this equilibrium is denoted by $B(E_c{}'(x_3,y_3))$ where $B(E_c{}') = \{(x,z) \in \Re_+{}^2 : x > x_3, y > y_3\}$

Proof: The proof is the same as in theorem (3).

Proposition 3. The equilibrium $E_c{}'(x_3,y_3,0)$ is stable with condition $0 > 1 - \dfrac{a\alpha(1-x_3)}{b(h_1 - h_2)}$

Proposition 4. The stability near the equilibrium $E_c{}^*(x^*,y^*,z^*)$ is given by equation

$$\lambda^3 + A\lambda^2 + B\lambda + C = 0 \text{ where}$$

$$A = ax^* > 0, \; B = bx^*c(y^* + z^*) + \alpha^2 y^* z^*, \; C = a\alpha^2 x^* y^* z^* > 0$$

$$AB - C - = ax^*(bx^*c(y^* + z^*)) > 0$$

From Routh-Hurwitz stability criterion it is stable.

4. Conclusion

In this paper, the discussion and analysis model prey predator interaction with harvesting of predator is presented. Boundednessof solution, and equilibriums points with their conditions were discussed. The basin attraction of some of equilibrium points was also calculated. Finally, the result shows us the infected predator increases while the susceptible predator decreasing.

REFERENCES

[1]   Azar, C., J. Holmberg, et al. Stability analysis of harvesting in a predator-prey model. Journal of Theoretical Biology., 174(1) (1995), 13-19.

[2]   Chevé, M., R. Congar, et al. (2010). Resilience and stability of harvested predator-prey systems to infectious diseases in the predator.

[3]   Dubey, B., P. Chandra, et al. A model for fishery resource with reserve area.Nonlinear Analysis: Real World Applications., 4(4)(2003), 625-637.

[9]   Haque, M. and E. Venturino.The role of transmissible diseases in the Holling–Tanner predator–prey model. Theoretical Population Biology., 70(3)(2006), 273-288.

[5]   S.A.Wuhaib,Y.AbuHasan, (2012).Apredator Infected Prey Model With Harvesting of Infected Prey .ICCEMS2012: 59-63

[6]   Venturino, E. Epidemics in predator–prey models: disease in the predators. Mathematical Medicine and Biology., 19(3)(2001), 185-205.

[7]   Y.X,L.Chen. Modelling and analysis of a predator prey model with disease in the prey. Mathematical Biosciences., 171(2001), 59-82.

www.ingramcontent.com/pod-product-compliance
Lightning Source LLC
Chambersburg PA
CBHW081555220326
41598CB00036B/6679